Multiscaled PVA Bionanocomposite Films

Mohanad Mousa · Yu Dong

Multiscaled PVA Bionanocomposite Films

Characterisation and Nanoscale Modelling

Springer

Mohanad Mousa
Shatra Technical Institute
Southern Technical University
Basra, Iraq

Yu Dong
School of Civil and Mechanical Engineering
Curtin University
Perth, WA, Australia

ISBN 978-981-15-8770-2 ISBN 978-981-15-8771-9 (eBook)
https://doi.org/10.1007/978-981-15-8771-9

This Springer imprint is published by the registered company Springer Nature Singapore Pte Ltd.
The registered company address is: 152 Beach Road, #21-01/04 Gateway East, Singapore 189721, Singapore

Preface

Biopolymers have gained significant attention due to an ever-increasing level of synthetic polymers accumulated in a different environment. Polyvinyl alcohol (PVA) is considered as one of popular water-soluble and ecofriendly biopolymers due to its recyclability and biotribological properties for widespread biotechnological applications. Its key drawback lies in insufficient thermal and mechanical properties when compared with those of synthetic polymers. Different nanofillers such as carbon nanotubes (CNTs), graphene oxides (GOs) and nanoclays are generally incorporated into PVA matrices in a nanocomposite system in order to improve the overall material performance. Carbon-based nanofillers like CNTs, carbon black and GOs are proven to be effective reinforcements because of unique structures and properties despite their nanotoxicity and high material cost. Bamboo charcoals (BCs) are ecofriendly and sustainable carbon-based particles in possession of good affinity with PVA to achieve excellent properties of PVA/BC bionanocomposites. Their porous structures facilitate the penetration of polymeric molecules resulting in strong filler-matrix interfacial bonding.

This book covers the holistic development of a new ecofriendly bionanocomposite system based on PVA and BCs as biopolymer matrices and nanofillers, respectively and investigates the possibility to take BCs as cheap, widely abundant and sustainable non-toxic alternatives to conventional carbon based nanofillers in comparison with clay-based counterparts including montmorillonite (MMT) and halloysite nanotubes (HNTs). A simple solvent casting method was employed to prepare novel and strong PVA/BC bionanocomposites, PVA/Cloisite 30B clay bionanocomposites, as well as PVA/HNT bionanocomposites. The multifaceted effects of different nanofiller sizes, contents, shapes and structures on their overall material performance were evaluated in a systematic manner. More importantly, a novel approach to experimentally measure 3D interphase dimensions and properties between different nanofillers and PVA matrices was elaborated via peak force quantitative nanomechanical tapping mode (PFQNM). Such important interphase features were utilised in developing a theoretical modelling framework to estimate effective volume fraction with respect to interphase volume and volume fraction, which was integrated into Halpin-Tsai model and Mori-Tanaka model in composite theory and achieved much better agreement with corresponding experimental data in higher accuracy.

In this book, Chap. 1 gives a detailed literature survey about biopolymers and bionanocomposites with a particular focus on PVA based bionanocomposites reinforced with BCs, MMTs and HNTs. This chapter also demonstrates nanocomposite fabrication, different properties like mechanical and thermal properties, as well as interphase nanomechanical properties potentially used for micromechanical modelling. Chapter 2 reports the material selection of PVA, BCs, Cloisite 30B clays and HNTs, material fabrication and experimental characterisation techniques employed according to the selected material formulation. Chapter 3 describes the preliminary assessment to detect material characteristics of BCs in both micro- and nanosizes. The influences of BC particle sizes and contents on material characterisation, mechanical and thermal properties, as well as nanomechanical properties of PVA/BC bionanocompisite films are revealed in order to manufacture new eco-friendly and super strong PVA/BC bionanocomposite films. A direct experimental measurement of Young's modulus of BCs via atomic force microscopy (AFM) is also reported for the first time. Chapter 4 presents a holistic study in relation to the effect of different nanofiller shapes, structures and contents on material characterisation of bionanocomposite films where Cloisite 30B clays, HNTs and BCs are deemed as typical 1D platelet-like, 2D tubular and 3D irregular-shape nanofillers, respectively. This chapter uses most of content materials from our published research article "*Mousa M, Dong Y (2020) "The role of nanoparticle shape and structures in materials characterisation of polyvinyl alcohol (PVA) bionanocomposite films" Polymers 12(2): 264*" under Creative Commons Attribution License. Chapter 5 introduces a novel approach to measure 3D interphase properties and dimensions in bionanocomposite systems in terms of interphase surface area and interphase volume. Chapter 6 illustrates a micromechanical modelling work to predict the elastic modulus of PVA based bionanocomposites with the estimated results being compared with those obtained from experimental data. Moreover, micromechanical models were further developed to consider the incorporation of 3D interphases in bionanocomposite systems in relation to interphase surface area and interphase volume. Both Chaps. 5 and 6 are based on our published research article "*Mousa M, Dong Y (2020) Towards sophisticated 3D interphase modelling of advanced bionanocomposites via atomic force microscopy, J Nanomater, Article ID* 4526108" under Creative Commons Attribution License. Chapter 7 reviews the applications of bionanocomposite materials with a particular emphasis laid upon electronic and sensor applications, as well as medical applications, along with the specific recommendations in relation to developed bionanocomposites in this study. Finally, Appendix provides the supporting information on "Interphase Dimension Measurement via AFM".

The first author Mohanad Mousa would like to express the gratitude to the Higher Committee for Education Development (HCED) in Iraq to offer the Ph.D. scholarship at Curtin University, Perth, Australia to conduct the research where the relevant results are disseminated in this book. We also sincerely acknowledge continued technical supports received from Mr. David Collier, Mr. Graeme Watson, and Mr. Andreas Viereckl at Mechanical Engineering laboratories, Ms. Elaine Miller and Ms. Veronica Avery at John de Laeter Centre, as well as Dr. Thomas Becker at Nanochemistry Research Institute, Curtin University. In particular, the success of this published book cannot be achieved without the editorial team at Springer Nature including Loyola D'Silva, Kavitha Palanisamy and N. S. Pandian.

We welcome valuable comments and feedbacks from other researchers, scientists and materials engineers to improve the overall quality of this book. We also wish that it can provide an essential guidance to sustainable bionanocomposite systems from material processing, experimental characterisation to modelling work in a holistic manner.

Perth, Australia

Mohanad Mousa
Yu Dong

Contents

Chapter 1
Introduction to PVA-Based Bionanocomposite Films

Abstract Biopolymers have gained their great popularity over synthetic polymers since conventional non-recyclable polymeric wastes are easily accumulated to negatively impact environmental sustainability and ecological systems. Polyvinyl alcohol (PVA), in possession of some favourable characteristics of good recyclability and biotribological properties, can be widely used in biotechnological applications, which is deemed as one of popular water-soluble and ecofriendly biopolymers. Nonetheless, it is also worth mentioning that insufficient thermal and mechanical properties of such biopolymers inevitably make them less competitive as opposed to commercially available synthetic polymers. Such material demerits of PVA can be overcome by reinforcing a variety of nanoreinforcements consisting of carbon nanotubes (CNTs), halloysite nanotubes (HNTs) and nanoclays in order to generate new nanocomposite systems for enhancing overall material performance. This chapter elaborates preparation, characterisation and properties of different PVA bionanocomposites reinforced with layered silicates, HNTs and BCs, as well as other popular fillers to gain the insight of their wide range of applications.

Keywords Bionanocomposites · Polyvinyl alcohol (PVA) · Biopolymers · Nanofillers

1.1 Introduction

Advances in material science have yielded an increasing annual production of synthetic polymers. To date, 8.3 billion metric tons of synthetic polymers have been produced [1], among which 6.3 billion metric tons becomes typical wastes. In between, about 9% have been recycled as opposed to 12% incinerated wastes. The other 79% plastic wastes are stored in landfills or directly released into the environment [1]. The majority of such plastics are petroleum-based synthetic polymers such as polypropylene (PP), polyethylene (PE) and polystyrene (PS) [2–5]. For example, the sole annual productions of PE and PP have reached more than 70 and 50 million metric tons, respectively [6]. Such polymers are generally regarded as non-degradable wastes after the consumption, and are often disposed of directly in the environment to cause severe pollution like dumped solid wastes and marine plastic

M. Mousa and Y. Dong, *Multiscaled PVA Bionanocomposite Films*,
https://doi.org/10.1007/978-981-15-8771-9_1

pollution. It has been reported that about 5.25 trillion plastic particles (weighing 269,000 tons) are floating on the sea [7]. such ubiquitous plastic particles in form of micro- or nanosizes in the marine environment could result in the entanglement of species by marine debris including marine mammals and other species in net fragment litter [7].

Moreover, proposed solutions such as waste accumulation in landfill are unsatisfactory on the long term because of limited landfill sites in continuous urban development. In addition, the incineration of plastic wastes produces a large amount of carbon dioxide and hazardous emission such as dioxins, resulting in the detrimental effect on our ecosystems. On the other hand, recycling plastic wastes is often non-economical with the availability of insufficient processing infrastructures while the quality of recycled plastics is lower than that of virgin materials. In response to these concerns, the development of biopolymers has become an increasingly high priority for material scientists and engineers. Such polymers would not involve the use of toxic gases in their manufacture, and thus can easily degrade in the environment. According to different manufacturing routes, biodegradable polymers can be classified as (i) agro-polymers based on the biomass of agro-resources as raw materials (e.g. starch and cellulose), (ii) polymers based on microbial production (e.g. polyhydroxyalkanoates), (iii) polymers that are chemically synthesised using biomass-based monomers (e.g. poly(lactic acid) (PLA)) and (iv) polymers that are produced from the chemical synthesis of petroleum resources [4, 8]. Among those categories, polyvinyl alcohol (PVA), as one of most popular water-soluble polymers, has practically gained considerable attention due to its favourable mechanical properties, thermal resistance, excellent flexibility [9], recyclability and biotribological properties [10]. Furthermore, its biotechnological applications comprise tissue engineering, drug delivery, articular cartilage and biosensors [11].

In recent years, the incorporation of nanofillers such as montmorillonite (MMT) clays [12, 13], halloysite nanotubes (HNTs) [14–16], carbon nanotubes (CNTs) [17, 18], graphene sheets [19, 20], cellulose nanocrystals (CNCs) [21, 22], laponite [23] and nanodiamond [24] into PVA matrices in newly developed nanocomposite systems have be the major focus in order to significantly improve mechanical and thermal properties, as well as biodegradability of PVA nanocomposites.

CNTs are considered as effective reinforcements widely used in polymer nanocomposites. In addition to their high material cost, nanotoxicity is deemed as the significant drawback of CNTs due of their accumulation in cytoplasm with resulting ability to destroy human cells under certain inhalational conditions [25]. On the other hand, graphene sheets have also experienced limited use owing to their tendency to form agglomerates [26]. More recently, BCs, as newly used carbon-based particles, are employed as effective reinforcements with unique ecofriendly and environmentally sustainable material features. Such BC particles are generally produced from carbonised bamboo and bamboo residues at a typical temperature of 1000 °C under nitrogen atmosphere [27], which results in the formation of significant amounts of lengthwise and crosswise pores within BC structures [27, 28]. More impressively, their volumetric porosity, mineral constituents and absorption efficiency are approximately 5, 8 and 10 times more than those of wood charcoals [27]. Besides, inner surface areas of BCs are typically in range of 250–390 m^2/g as opposed to 10 m^2/g for wood charcoals [29, 30]. Due to aforementioned remarkable characteristics, the use of BCs has led to widespread applications such as negative ion supply [31], humidity regulators [32], water purification [33], oxidation prevention, antibacterial [34] and antifungal features, as well as breathability [30]. The ability of BCs to regulate humidity directly benefits the hindrance to the growth of bacteria and fungi with well-maintained freshness of food [34], which is a key requirement for food packaging industries in order to increase the shelf life of products. Additionally, BCs are also capable of absorbing significant levels of infrared energies from the environment, which can be then emitted to support cell activation for efficient human blood circulation [35]. Elemental inclusions within BCs such as calcium, potassium, sodium and iron have also been utilised for the purposes of food cooking, baking and storage [36, 37].

Apart from BCs, several types of clays have been used in nanocomposite systems like kaolin [38], mica [39], sepiolite [40] and MMT [12]. Among them, MMT is most popular clay fillers owing to their layered silicate structures with exchangeable cations and reactive –OH groups on MMT layer surfaces [25], as well as high strength and stiffness, large aspect ratio with natural abundance and availability in large quantities [25, 41]. MMT is capable of forming stable suspensions in water while its hydrophilic characteristic also promotes the dispersion of these inorganic crystalline layers in water-soluble polymers like PVA [12]. Moreover, layered silicates in their pristine state has high ion-exchange capacity, which allows for the modification of interlayer spacing in order to achieve better compatibility with base polymers [25]. Modified MMTs such as Cloisite 30B clays, containing 90 meq/100 g of a methyl, tallow, bis-2-hydroxyethyl ammonium chloride surfactant [42], has been used in many studies for the preparation of bionanocomposites. For instance, Cloisite 30B clays contain –OH groups to be more compatible with base polymers when compared with Na^+ MMTs and Cloisite 15A counterparts [43]. Moreover, Cloisite 30B clays have much larger *d*-spacing value as opposed to those of other modified clays like Cloisite 15A, Cloisite 20A and Cloisite 25A clays [44].

HNTs, as the alternative aluminosilicate clays [$Al_2Si_2O_5(OH)_4 \cdot nH_2O$], has received great attention owing to their novel 2D natural hollow nanotubular structures, resembling that of multi-walled CNTs (MWNTs) [45]. In addition, HNTs have high aspect ratios, good functionality and biocompatibility, as well as high mechanical strength [45]. These aforementioned material characteristics warrant the wide use of HNTs in biological/medical applications such as bioreactors, drug delivery and nanocoating [45]. Moreover, HNT surfaces contain alumina and silica groups to facilitate hydrogen bonding interactions between HNTs and PVA matrices, which makes HNTs an ideal nanofiller candidate for PVA bionanocomposite films [45].

1.2 Biopolymers

Biodegradable polymers or biopolymers are defined as those polymeric materials that can degrade in a bioactive environment via the enzymatic action of microorganisms such as bacteria, fungi and algae. Moreover, some polymeric chains may also undergo a scission-down by nonenzymatic processes like chemical hydrolysis. Among different types of biopolymers, polylactide or polylactic acid (PLA) is regarded as one of the most widely produced bioplastics [46]. It is a linear thermoplastic polymer mainly derived from renewable resources such as corns or sugar beets [46]. Fully biodegradable and non-toxic PLA is produced from lactic acid that possesses a chiral molecule in the dextrorotatory form as L-(+)-lactic acid or else in the levorotatory form as D-(-)-lactic acid. The insufficient properties of some PLA polymers in relation to thermal stability, impact resistance and flexibility can be improved by their blending with other polymers like poly(ethylene oxide) (PEO) or by using a plasticiser such as oligomer lactic acid (OLA), citrate ester or low-molecular-weight polyethylene glycol (PEG) [8, 47]. Moreover, it is worth noting that the processing temperature of PLA should exceed its thermal degradation temperature taking place at 200 °C. However, the microbial degradation of PLA seems still challenging because PLA requires the temperature level above its glass transition temperature (T_g) in range of 55–62 °C for the onset of PLA hydrolysis, which makes its degradation difficult at ambient temperatures. Additionally, the degradation rate of PLA depends on its molecular weight. In general, high-molecular-weight PLA is mechanically stronger but less susceptible to biodegradation [48].

Polyhydroxyalkanoates (PHA) as another biopolymer belongs to a family of naturally occurring hydrophobic, biocompatible and biodegradable polyesters [49, 50]. It is available in a wide variety of forms, and used for carbon or energy storage in microorganism in form of light refracting granules inside the cells. According to their chain length, PHA can be divided into three major types, namely short chain length (scl PHA with a carbon-number range of C3–C5), medium chain length (msl PHA with the corresponding range of C6–C14) and long chain length (lcl PHA with carbon number greater than C14) [51]. Polyhydroxybutyrate (PHB) and poly(hydroxybutyrate-cohydroxyvalerate) (PHBV) are the two most common PHA-based polymers with a wide range of microorganisms used to accumulate PHB. However, the genetic manipulation is essential for the large-scaled production of PHB, whereas alcaligenes eutrophus is the most common microorganism implemented in the biosynthesis of PHA [47, 49–51]. PHB is comparable to polypropylene (PP) due to several similar physical properties such as melting point, degree of crystallinity and T_g. Nevertheless, PHA exhibits higher stiffness, lower toughness, lower solvent resistance and higher natural resistance to ultraviolet radiation when compared with PP [49]. The main disadvantage of PHA lies in its narrow processing window since it starts to degrade under the condition of high shear and high temperature.

Poly(ε-caprolactone) (PCL) is a linear polyester synthesised via the ring-opening polymerisation of ε-caprolactone in the presence of metal alkoxides [2]. It possesses good flexibility and high elongation at break, which makes it attractive for various applications such as polyvinyl chloride (PVC) plasticisation, drug release control and soft compostable packaging [2]. The main drawback of PCL is its low melting point occurring at 65 °C, which can be alleviated by polymer blending [52, 53] or by modifications such as cross-linking [54]. PCL biodegradability can be assessed in the presence of fungi, which is ready for enzymatic degradation [54]. Such similar effect is also achieved in bacteria and yeast [55]. The low hydrolysis rate of PCL homopolymer [56, 57] has been shown to be improved significantly when blended with starch [52]. However, high production cost is considered as its main material demerit.

In contrast to above-mentioned biopolymers, water-soluble polymers have a broad range of applications such as food and pharmaceutical use, detergent builders, scale inhibitors, adhesives and so on [58]. A majority of water-soluble polymers are prepared from acrylic acid, maleic anhydride, methacrylic acid and various combinations of these monomers. Except their oligomers, these polymers are commonly non-biodegradable. Therefore, conventional water-soluble polymers persist in water depositories such as oceans and lakes and their effects are harder to be recognised by consumers with potential problems [58]. Water-soluble biopolymers could be synthesised via the modification of starch and cellulose such as carboxymethyl cellulose (CMC) and hydroxyethyl cellulose (HEC). However, the biodegradability of such polysaccharide-derived polymer decreases when a high level of cellulose modification is used. Moreover, many of these polymers have not yet been studied by standard testing methods to determine their biodegradability [58].

Poly(amino acids) with free carboxylic groups, such as poly(aspartic acid) and poly-(glutamic acid), are considered as suitable material candidates for water-soluble biopolymers. As such, those polymers based on aspartic acid can offer greater commercial success. Poly(aspartic acid)s (thermal polyaspartate, TPAs) are functionally equivalent to poly(acrylic acid), and become fully biodegradable when they are highly linear in chemical structures [58, 59]. TPAs used in a wide range of applications such as performance chemicals, diapers and agriculture. Poly(malic acid)s, the polyester equivalent of poly(aspartic acid)s, may be useful polymers for biodegradable detergents despite their hydrolytically unstable characteristic [58].

To date, PVA is the only water-soluble polymer with exclusively carbon atoms in main chains with good biodegradability. It is currently used in textiles, paper and packaging industries as paper coatings, adhesives and films [58].

1.3 Poly(Vinyl Alcohol) (PVA)

PVA is a water-soluble polymer with 1, 3-diol units or 1, 2-diol units, which depends on the hydrolysis of poly(vinyl acetate). In addition, the content of 1, 2-diol units is reduced when the temperature of vinyl acetate polymerisation decreases. The

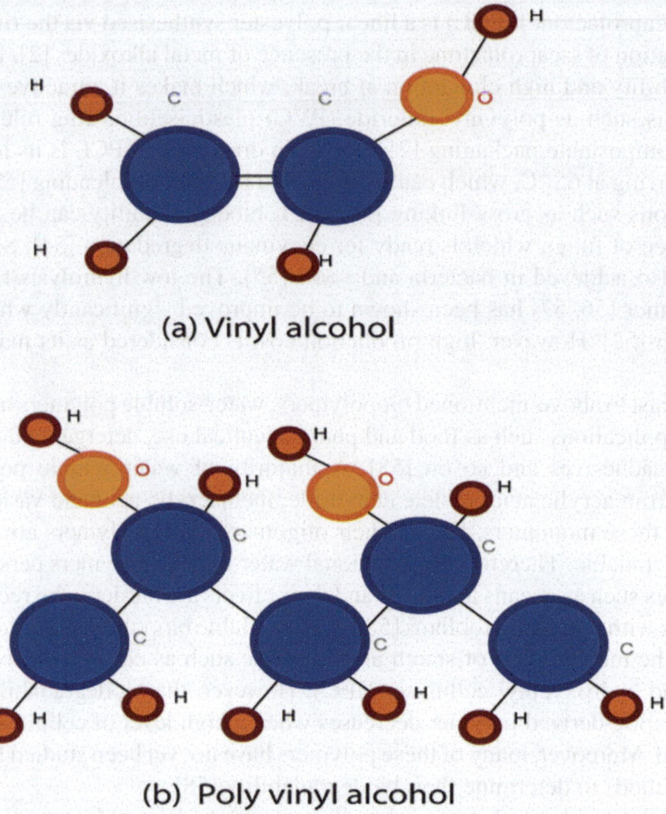

Fig. 1.1 Chemical structures: **a** viny alcohol and **b** PVA [25]

chemical structures of PVA and viny alcohol are shown in Fig. 1.1. PVA has been synthesised by several routes such as the polyaldol condensation of acetaldehyde despite its little success until now [60]. In addition, the polymerisation of metal compounds, followed by the saponification process, produces low-molecular-weight PVA. However, the synthesis of PVA by the polymerisation of vinyl ester or ethers, further subjected to the saponification or esterification, is regarded as the most popular method [13, 60, 61]. In this technique, vinyl ester or other derivatives such as vinyl mono and dichloroacetate are usually used as base materials for manufacturing PVA. Moreover, PVA characteristics like molecular weight and residual content of acetyl groups can be controlled by optimising processing parameters such as polymerisation temperature, vinyl acetate/methoanol ratio and polymerisation conversion [62]. It has been reported that the decrease in polymerisation temperature with an appropriate proportion of methonal and conversion yields an increase in the molecular weight of final PVA [62]. In addition, PVA should be free from monomeric vinyl acetate to avoid the cosaponification [60]. The conversion process of PVAc to PVA is usually carried out in the solution, suspension, or emulsion in the presence of alkaline or acidic catalysts. As such, the most recommended process is the transesterification in the methanol using a proper amount of sodium methoxide as the catalysts. The residual content of acetyl groups can be controlled by manipulating catalyst content, reaction temperature and reaction time. Moreover, the types of catalysts and solvents used are selected based on the distribution nature of acetyl groups in the partial saponification of PVA. For example, acetyl groups are distributed in a blockwise manner in case of alkaline saponification while they distributed statistically with respect to acidic saponification [62].

PVA properties are determined from molecular weight and degree of hydrolysis, as depicted in Fig. 1.2. When these two parameters increase, water resistance, tensile strength, block resistance and solvent resistance are enhanced while solubility, flexibility and water sensitivity diminish accordingly [63]. Remarkable PVA properties such as its resistance against organic solvents, water solubility (even in cold water), low cost, good mechanical properties and excellent barrier properties [58] make it an excellent biopolymer candidate for various applications ranging from food packaging, textile products, paper coating, finishing adhesives to medical devices [13, 61, 63]. The biocompatibility of PVA has been widely investigated with its apparent non-toxic characteristic. For example, it has been reported that subcutaneous and intramuscular implantation of PVA hydrogel into rabbits did not induce any adverse effects on the surrounding tissues [61]. Similarly, the injection of PVA hydrogel into the eyes of crab-eating macaques did not show any evidence, even after three-month

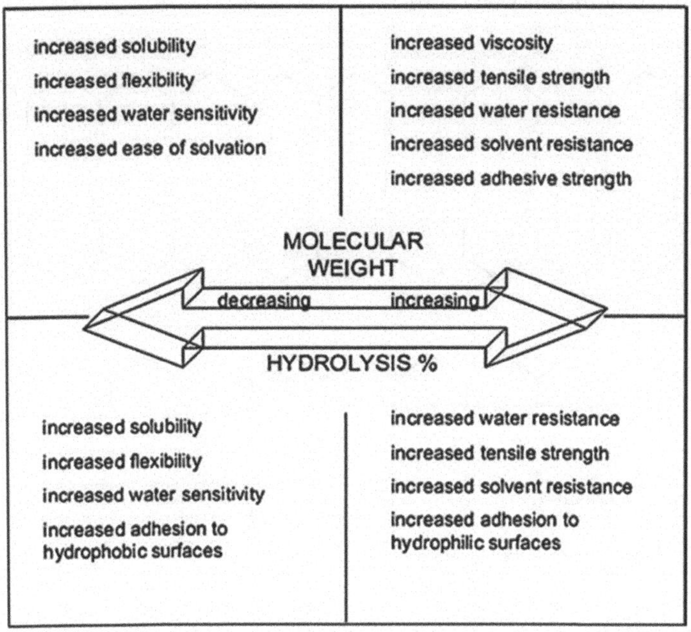

Fig. 1.2 Effect of degree of hydrolysis and molecular weight on PVA properties [25]

tissue loss, as compared to the change in opthalmoscopic finding or the increase in intraocular pressure [58]. In addition, after the investigation of PVA biocompatibility by food industries, PVA is recommended to be used for the manufacture of safe biopolymer products [13, 61, 63].

In general, PVA is used after being dissolved in water. Such a polymer can be dissolved in water when the entire strength of solute-solvent hydrogen bonding exceeds that of PVA inter-molecular hydrogen bonding [64]. As a result, PVA solubility in water is mainly determined by the degree of hydrolysis. It is noted that the solubility of PVA with a partial degree of hydrolysis is higher than that of full-hydrolysis crystalline counterpart [64]. This could be attributed to strong intermolecular hydrogen bonding in case of full hydrolysis when compared with partial hydrolysis, as shown in Fig. 1.3 [65]. To ensure that PVA has been completely dissolved in water, the preparation temperature must exceed 80 °C over an acceptable time period [65]. It is believed that the inter- and intrachain hydrogen bonding of PVA is disrupted by thermal energy. Moreover, the interaction between PVA chains leads to the increase in solution viscosity, which may further result in phase separation, depending on its degree of hydrolysis, solid content and storage modulus [66, 67]. On the other hand, partial-hydrolysis PVA can be dissolved in water at room temperature more readily than full-hydrolysis counterpart [68]. Moreover, the viscosity of PVA demonstrates a marginal change with the storage time, which is ascribed to more acetyl groups in PVA with the partial hydrolysis to be responsible for disrupting inter- and intrachain hydrogen bondings, thus resulting in the increase of polymeric solubility in water [68].

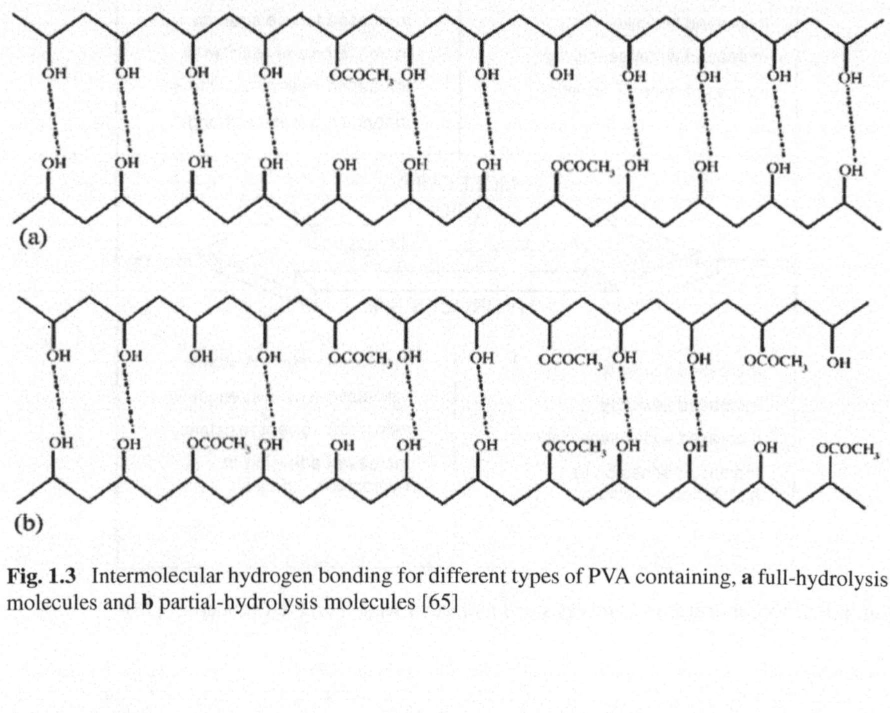

Fig. 1.3 Intermolecular hydrogen bonding for different types of PVA containing, **a** full-hydrolysis molecules and **b** partial-hydrolysis molecules [65]

The molecular weight (MW) of PVA is also considered as an important factor in controlling solubility. In case of full-hydrolysis PVA, the solubility decreases with increasing the molecular weight, which is associated with the increase in PVA molecular weight leading to the higher possibility of intramolecular hydrogen bonding between hydroxyl groups in side branches [68]. As such, these types of bonds make PVA difficult to dissolve in water. For example, Chang et al. [65] noted that full- hydrolysis PVA with MW = 22,000 dalton (Da) was higher than that of full-hydrolysis PVA with MW = 74,800 dalton (Da). However, partial-hydrolysis PVA demonstrates different solubility behaviour. The solubility of PVA increases with increasing its molecular weight, which is attributed to a small number of hydroxyl groups in partial-hydrolysis PVA, and this is insufficient to build up the intramolecular hydrogen bonding [68]. In addition to molecular-weight effect, electrolyte impact on hydrogen bonding in a water-soluble system has also been investigated [69]. It is believed that the use of electrolyte can disrupt hydrogen bonding between polymeric chains with an decrease in the viscosity of polymer solutions [69]. In case of PVA, the additional NaCl can induce the disruption of inter- and intrachains for hydrogen bonding. That is why it is essential to enhance both PVA solubility and viscosity in water. Nevertheless, the further addition of NaCl leads to the disruption of solute-solvent hydrogen bonding with decreasing the solution viscosity [70].

1.4 Nanofillers for Bionanocomposites

1.4.1 Layered Silicates

Natural or synthetic mineral-layered silicates have gained great popularity as effective reinforcements for nanocomposite systems. These types of fillers comprise the stacks of layers separated by Van der Waals interactions with gallery spacing. Each layer is typically 1 nm thick and 30 nm to several microns long [71, 72]. Depending on clay sources, silicate types and synthesis techniques, these layers can be bound together with counter ions. Each layer comprises a tetrahedral sheet containing a silicon atom surrounded by four oxygen atoms and an octahedral sheet where metals such as aluminium or magnesium hydroxide are surrounded by eight oxygen atoms. For instance, kaolinite 1:1 layered structure consists of a silicon tetrahedral sheet and aluminium octahedral sheet that share a common plane of oxygen atoms [71]. In comparison to this, the crystal structure for 2:1 phyllosilicates has two tetrahedral sheets with a central octahedral sheet of alumina (Fig. 1.4) [73, 74]. Pyrophyllite is generated in a 2:1 crystal structure containing silicon in tetrahedral sheets and aluminium in an octahedral sheet without any partial or complete replacement of atoms. Pyrophyllite layers do not expand in water and possess only outer surface areas. When the substitution takes place between the silicon in tetrahedral sheets and aluminium in octahedral sheets, the final structure is known as mica. Any negative hole resulting from mineral substitution is counter balanced by potassium cations

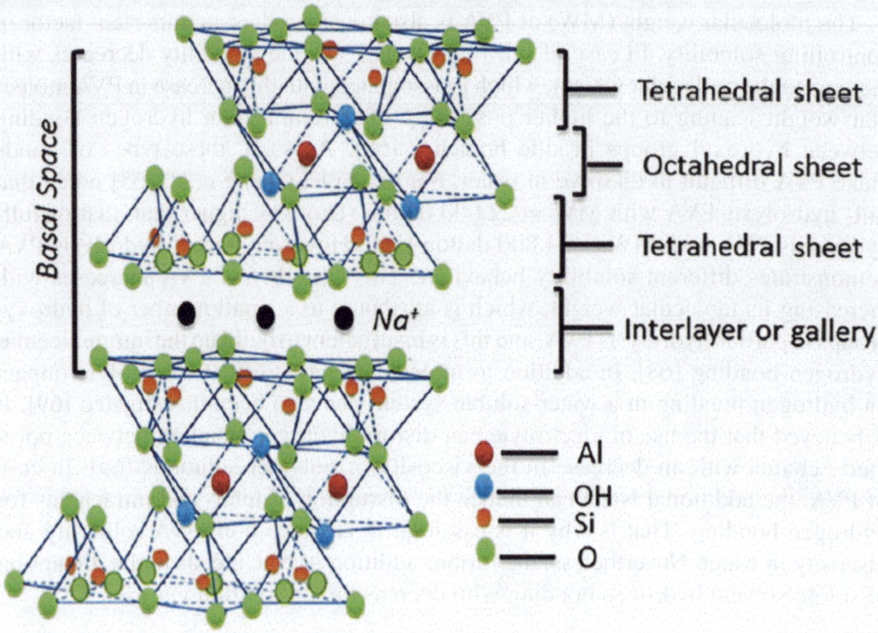

Fig. 1.4 Crystal structure of 2:1 layered silicates [75]

located in the gallery. Mica has a similar behaviour to pyrophyllite in that it does not expand in water and only possesses external surface areas. This is attributed to the potassium cation occupying the holes resulting from Si/Al replacement. On the other hand, a partial substitution between Al^{3+} and Mg^{2+} is required in order to ensure that layered silicates possess the ability to swell in water and hydrophilic polymers. Such a clay type is known as montmorillonite (MMT). Additionally, the negative charge is balanced by sodium or calcium ions that fit the gap leading to the weak bonding between the layers. Thus, it clearly explains why MMTs are commonly used as fillers in polymer nanocomposites. Apart from MMTs, hectorite and sponite are also attractive fillers owing to their high aspect ratios and good dispersibility, as well as their easy surface modification using organic or inorganic surface cations [74, 76]. Chemical formulae for these nanoclays are given in Table 1.1 The pristine form of layered silicates containing sodium and calcium ions alone is only miscible with hydrophilic polymers such as polyvinylpyrrolidone (PVP) [77] and PEO [78].

Table 1.1 Chemical structures of commonly used 2:1 phyllosilicates [76]

2:1 Phyllosilicate	General formula
Montmorillonite	$M_x\ (Al_{4-x}Mg_x)\ Si_8O_{20}(OH)_4$
Hectorite	$M_x\ (Mg_{6-x}Li_x)\ Si_8O_{20}(OH)_4$
Saponite	$M_x\ Mg_6(Si_{8-x}Al_x)\ O_{20}(OH)_4$

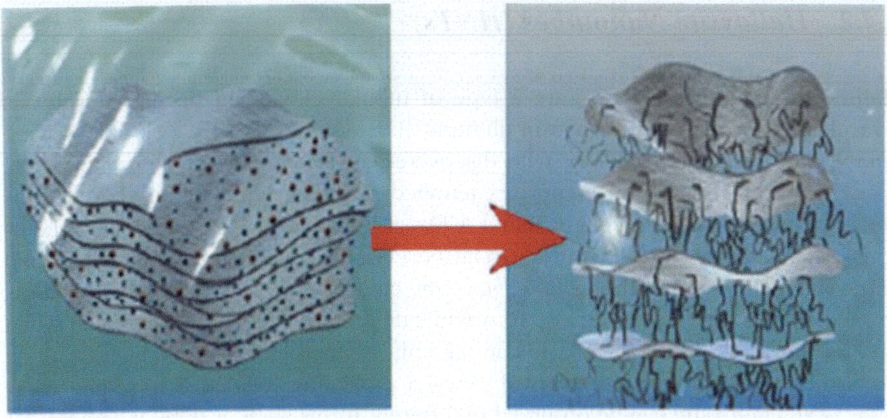

Fig. 1.5 Schematic diagram of the modification of layered silicates by ion exchange reactions [81]

In order to make nanoclays miscible with hydrophobic polymers, their surface condition must be modified from hydrophilic to organophilic feature using ion-exchange reactions according to a typical method where alkali-cation ions are substituted for organic compounds like alkylammonium ions, sulfonium and phosphonium (Fig. 1.5) [78, 79].

These organic surfactants can minimise surface energy of inorganic host and enhance clay wettability by using hydrophobic polymers with increasing the interlayer spacing. In addition, functional groups of organic compounds have also been utilised to improve the bonding between layered silicates and polymers [80, 81]. The chain length of surfactants and charge density of clays are known to play an important role in controlling the interlayer spacing between clays. In general, the use of a long molecular chain in combination with high charge density tends to increase the interlayer spacing (Fig. 1.6) [79, 82].

Wang et al. [83] prepared organoclays with different chain lengths of alkylammonium. An increase in interlayer spacing values was achieved with increasing the length of alkylammonium. It was reported that the interlayer spacing of MMT modified with octadecylammonium was increased from 1.36 to 2.47 nm.

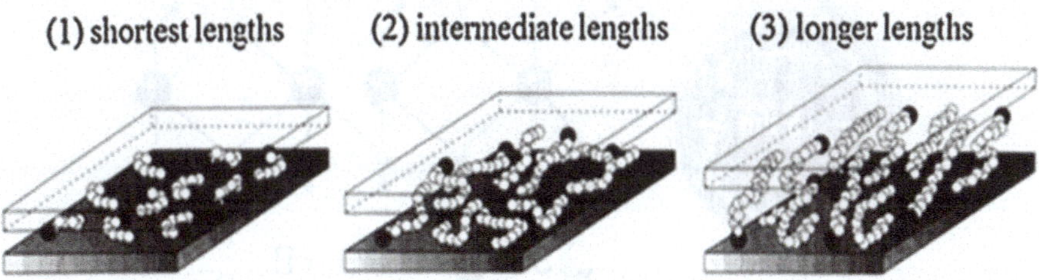

Fig. 1.6 Alkylammonium chain aggregation models [76]

1.4.2 Halloysite Nanotubes (HNTs)

Halloysite nanotubes (HNTs) are a type of tubular clays that occur naturally as a result of hydrothermal change in aluminosilicate minerals [45, 84]. HNTs were named after Omalius d'Halloy who discovered such a mineral material in 1826. The HNT crystal structure comprises tetrahedral sheets with corner-shared SiO_4 and octahedral sheets with edge-shared AlO_6, as depicted in Fig. 1.7. HNTs (chemical formula: $Al_2Si_2O_5(OH)_4 \cdot nH_2O$) can be classified as hydrated or dehydrated halloysite depending on the value of n. In the case of $n = 2$, hydrated halloysite is formed with an interlayer space of 10 Å while dehydrated halloysite is created when $n = 0$ [85, 86]. A second scenario can take place when HNTs are heated in range of 30–110 °C, leading to the removal of water molecules located in their interlayer areas [45]. Halloysite can be observed in different forms such as tubular, spheroidal and platy halloysite depending on different geological and crystallisation conditions, among which the tubular form is the most common one [87]. HNT surfaces contain low hydroxyl groups in addition to siloxane groups (Si–O–Si) while the interlayer region contains intensive hydroxyl groups with Al–OH. The main feature of having

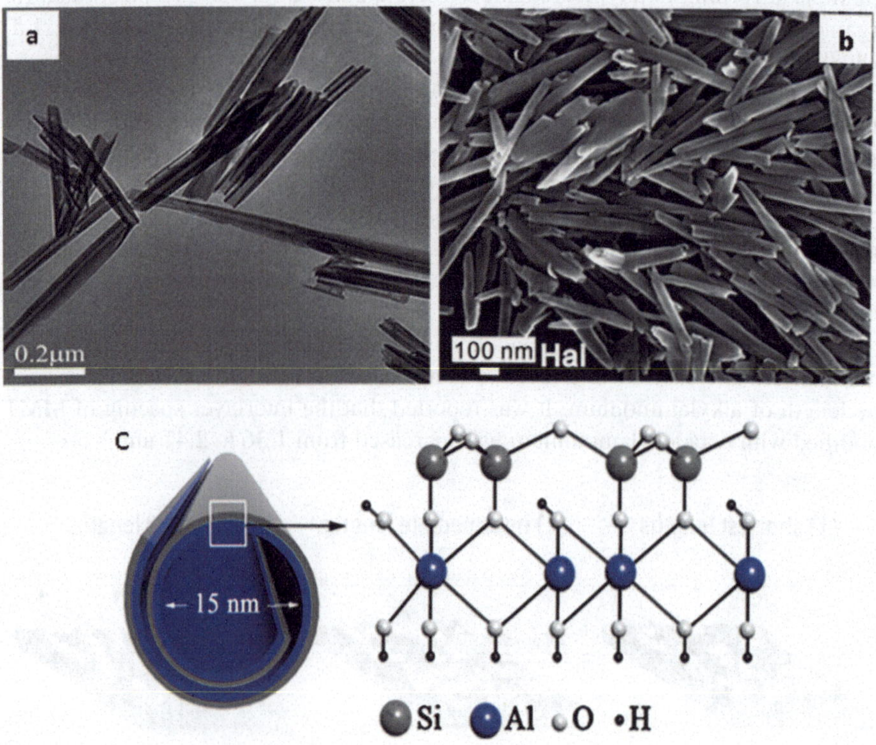

Fig. 1.7 Halloysite shape and structure in **a** a transmission electron micrograph and **b** a scanning electron micrograph, as well as **c** chemical structure of halloysite [25]

Table 1.2 Typical properties of HNTs [25]

Type of property	Description	Ref.
Chemical formula	$Al_2Si_2O_5(OH)_4 \cdot nH_2O$	[45, 85]
Aluminium (%)	20.9	[84]
Silicon (%)	21.76	[84]
Hydrogen (%)	1.56	[84]
Length	100–2000 nm	[87]
Internal diameter	30–50 nm	[87]
External diameter	1–30 nm	[87]
Aspect ratio	10–50	[45]
Elastic modulus (theoretical value)	140 GPa (230–340 GPa)	[88, 89]
Mean particle size in aqueous solution	143 nm	[90]
Particle size range in aqueous solution	50–400 nm	[90]
BET surface area	22.1–81.6 $m^2 g^{-1}$	[91]
Lumen space	11–39%	[45]
Density	2.14–2.59 g cm^{-3}	[45]
Structure water release temperature	400–600 °C	[45]

low numbers of hydroxyl groups on HNT surfaces is to minimise the tube–tube interaction, and thus to promote their miscibility with polymers or other solvents [92–95]. The main advantages of HNTs include their wide availability [45], cost-effectiveness, environmental friendliness [96], high aspect ratio [97] and high heat resistance in addition to their good mechanical properties, as shown in Table 1.2. HNTs are widely used for nanoreactors [87], sorbent for contaminants and drug delivery [85]. Most raw halloysite contains impurities such as illite, feldspar, perlite and metal ions to be removed before its use. However, previous studies [16, 98] have shown that pure HNTs can be obtained after stirring 10 wt% HNTs in distilled water with gradually improved HNT dispersion up to 60 °C. This step is followed by centrifugation and washing three times in distilled water before being dried in air at 60 °C for 12 h. Sodium hexametaphosphate [98] and Tween 80 [16] can be added to the suspension in order to improve the stability of HNTs within the suspension. In nanocomposite systems, HNTs are considered as promising nanoreinforcements due to their outstanding material features, demonstrating overall performance improvement of thermoplastics or their blends such as polyamide, PP/PCL and polybutylene terephthalate (PBT) via melt compounding without any modification [45, 87]. However, the surface functionalisation of HNTs is required to increase their compatibility with polymers in order to warrant maximum stress transfer from polymer matrices to HNTs in nanocomposites [45, 87]. Two approaches exist for HNT modification, namely covalent modification and non-covalent modification [45]. Among them, covalent modification uses the reaction between a covalent agent and hydroxyl groups on HNT surfaces to increase their interfacial adhesion.

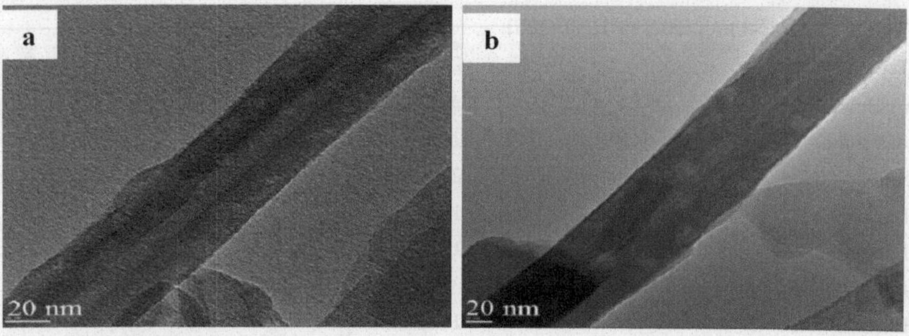

Fig. 1.8 Transmission electron micrographs of halloysite: **a** pristine HNT and **b** octadecylphosphonic acid-treated HNT [105]

This has been achieved by grafting HNTs with silane coupling agents such as γ-lycidoxypropyltrimethoxysilane (GPTS) [99], 3-aminopropyltrimethoxysilane (APS) [100], (3-aminopropyl) triethoxysilane (APTES) [101] or 3-(trimethoxysilyl)propyl methacrylate (MAPTS) [102] in the presence of toluene or water/alcohol mixture. Such a method can also be used as a pre-treatment before applying surface graft polymerisation as another HNT surface modification process [86]. The second method involves the grafting of HNTs by compatible polymeric chains such as poly(butylenes adipate) (PBA). It has been found that 3 wt% PBA-grafted HNTs could significantly increase the interfacial bonding between HNTs and polyvinylchloride [103]. The other study indicates that grafting HNTs with poly(methyl methacrylate) (PMMA) can increase the contact angle of water, decrease the polarity of HNTs and enhance toughness and wear resistance of resulting nanocomposites [104]. Phosphonic acid grafting is another method used for covalent modification, which can result in significant increases in interlayer spacing for HNTs (i.e. increasing from 7 to 15.1 Å), improved HNT dispersion and better performance for final nanocomposites [105]. Figure 1.8 illustrates an example of HNTs before and after modification. Though covalent modification gives rise to some improvements to nanocomposite performance, a relatively low number of hydroxyl groups existing on HNT surfaces offer limited reaction sites for covalent modification, which makes this method generally unsatisfactory for HNT modification [45]. Fortunately, the presence of metallic atoms such as Al and Fe without occupied orbital can render suitable opportunities to modify HNTs through an electron transfer interaction (i.e. non-covalent approach) [106]. During this process, chemical compounds with the ability to donate electrons such as 2,5-bis (2-benzoxazolyl) thiophene (BBT) [98] and 2,2-(1,2-ethene diyldi-4,1-phenylene) bisbenzoxazole (EPB) [107] are mechanically mixed with HNTs to form fibril structures within continuous polymer matrices.

This enhances the crystallinity level and mechanical performance of final nanocomposites [45]. Hydrogen bonding interaction is a second non-covalent modification technique in which organic compounds involving melamine (MEL), melamine cyanurate (MCA) or diphenylguanidine (DPG) are attached to HNTs by hydrogen bonding to create filler networks within polymer matrices [108]. It has been reported that HNT/MEL-based nanocomposites possess an increased flexural modulus of 3.65 GPa and flexural strength of 109.1 MPa when compared with those of unfilled polymer (1.93 GPa and 79.7 MPa, respectively) [109].

1.4.3 *Bamboo Charcoals (BCs)*

Bamboos, as naturally grown plants, are usually discovered in East Asia, but currently grow worldwide, especially in subtropical regions [110]. Moreover, bamboo forests cover approximately 22 million hectares, which is equivalent to 1% of total global forest areas [110]. Bamboos are also considered as inexpensive and ecofriendly materials without the required fertilisers, or pesticides for their growth. Bamboos can be converted into widely used charcoals with typical porous structures when carbonised or pyrolysed. The carbonisation is carried out at inert atmosphere and temperature above 700 °C when bamboos start to release the water content and organic gases, which is followed by the reconstruction of carbon structures [111]. Although the volume of bamboo can shrink significantly during the carbonisation, their tubular structures still well remain. Figure 1.9 presents morphological structures for the cross-sectional and lateral views of BCs carbonised at 1000 °C, which confirms apparent porous nature of original bamboos [110].

In previous work [112], the morphological structures and shrinkage of BCs were investigated at different carbonisation temperatures, which showed that the scaffolding of BCs was similar to that of original bamboos despite a slight change of bamboo microstructures during the shrinkage. Furthermore, the shrinkage ratios of bamboo carbonised at 500, 750 and 1000 °C were found to be about 21, 38 and 40%, respectively. Furthermore, morphological structures of bamboo charcoals under

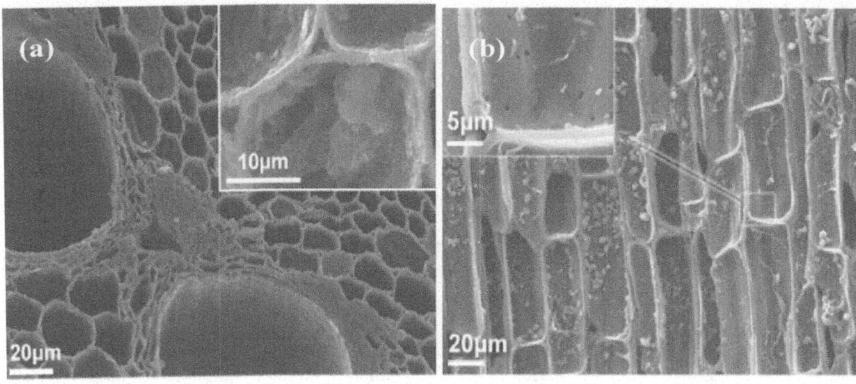

Fig. 1.9 Scanning electron micrograph of raw bamboo biotemplate after pyrolysis: **a** cross-sectional view and **b** lateral view [110]

pyrolysis at 1000–1500 °C possessed the same porous features of fresh bamboos, as well as exhibited a wide range of pore distributions from 1 nm to 1 μm in size [113]. On the other hand, rough walls of basic units inside BCs, namely parenchyma, along with their smooth outer surfaces can make entire charcoals very hard. This is because BCs consist of carbon with a small amount of minerals. The graphitisation behaviour of BCs treated in a temperature range of 2100–3000 °C was investigated by using X-ray diffraction (XRD) analysis in Fig. 1.10 [25, 114]. In general, XRD result for BC showed the existence of two broad bands at 22 and 43° corresponding to the C_{0002} and C_{0004} reflections of carbon, respectively. Moreover, it was clearly revealed that increasing the graphitisation temperature led to a decreasing spacing d_{002} of BCs while corresponding graphite crystallite size L_c increased instead.

The amount of minerals inside BCs is usually less than 10 wt%, and the amount and type of such minerals depend on the carbonisation temperature as well as geological vegetation source. For example, the bamboos 'phyllostachys edulis' obtained from Fujian Province, China, contain many additional elements such as Si, N, P, K, Mg, Al in addition to C, O and H [110]. As mentioned earlier, carbon represents the main element of BCs. This kind of biocarbon with special structures attracts the considerable attention to a wide range of applications such as the adsorption of harmful gases in air and the disposal of heavy metal ions in sewage water treatment [115]. BCs have high volumetric porosity and large surface areas of approximately 250–390 m^2/g as opposed to 10 m^2/g for wood charcoals. As such they can strongly absorb toxic gases in the environment [115]. Asada et al. [116] reported that the adsorption ability of BCs to benzene, toluene and skatole increased with increasing the carbonisation temperature. Additionally, BCs are also capable of emitting infrared waves that can be absorbed by human bodies. As a result, they can improve human microcirculation

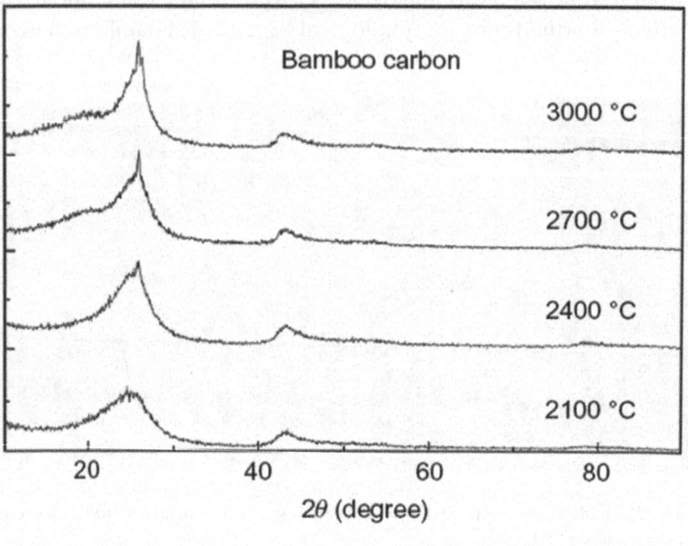

Fig. 1.10 XRD profiles of BCs after the carbonisation at different temperatures [25]

system, stimulate acupuncture meridians of bodies, promote metabolism and balance the body pH level and so on [115].

BCs, due to their carbon content, also possess certain electrical conductivity and magnetic properties [115]. Electromagnetic pollution becomes a critical issue owing to the excessive applications in electronic technologies. It has been reported that in a range of 10–1000 MHz, composite materials reinforced with BCs (thickness: 3 mm) have electromagnetic shielding effect of 45–75 dB [115]. Additionally, BCs have a special feature to prevent them from electromagnetic radiation, which can motivate manufacturers to develop man-made boards for IT applications. BCs are also considered as good materials in manufacturing supercapacitors and lithium-ion batteries [115]. In a nanocomposite system, BC morphology displays a wide range of pore distributions from less than 1 nm to 1 μm [110]. BC structure comprises parenchyma, which is the basic unit inside bamboo with the characteristic of rough wall surfaces, and with a large number of rough-walled pores inside BCs, polymeric chains tend to easily penetrate into internal BC pores [117] to form strong mechanical bonding in addition to the hydrogen bonding of pores. Hence, mechanical properties of resulting nanocomposites can be improved accordingly owing to the effective interfacial bonding between fillers and polymer matrices. Nevertheless, the selection of polymer matrices also plays a leading role in the preparation of nanocomposites in addition to BCs. Polymeric behaviour in a porous medium is associated with capillary forces, which enables to prevent polymeric chains from entering into internal pores [117]. According to the Laplace theory [117, 118], the strength of capillary forces can depend on surface chemistry and physical properties of polymers. When hydrophilic media are selected, applied positive capillary pressures drive polymeric chains into BC pores to form both mechanical and chemical bondings, which is why PVA has been chosen in this study. On the contrary, the surfaces of hydrophobic media generate negative capillary pressures and thus hinder the entry of polymeric chains into BC pores, leading to typical phase separation in a nanocomposite system. Such a mechanism can be interpreted as the debonding effect in polymer/BC composites including PLA/BC composites [31], ultrahigh molecular-weight polyethylene (UHMWPE)/BC composites [119] and polyaniline (PANI)/BC composites [120].

1.4.4 Other Popular Nanofillers

Other nanofillers worth mentioning include CNTs, nanocellulose and graphene oxides (GOs). CNTs are one of the most interesting carbon allotropes in the form of cylinders arranged by rolling graphene sheets [121], which results in the deformation of their sp^2 hybrid orbitals with the formation of a σ-π rehybridisation structure [121]. The main feature of the rehybridisation structure is that the confinement of π electrons offers unique and extraordinary properties of CNTs such as high Young's modulus at 1200 GPa and tensile strength at 150 GPa [122]. Enormous interest has been shown on CNTs owing to their diverse potential for supercapacitors [123], actuators in robotic manufacturing [124] and storage energy (e.g. hydrogen storage)

[125]. In a nanocomposites system, CNTs have the capability to improve mechanical and thermal properties, fire inflammability, moisture resistance and barrier properties due to their high-performance functionality [126]. It is worth noting that CNTs possess the ability to reinforce both thermosets and thermoplastics [127] in order to produce good mechanical properties and good affinity to chemical compounds [128]. Nevertheless, the critical issue currently being encountered is typical CNT agglomeration in the form of bundles when polymer/CNT nanocomposites are prepared. Such a phenomenon arises from the high aspect ratio of CNTs together with their strong Van der Waals interaction, particularly at the high CNT loadings.

Nanocellulose is the other popular nanofillers, which exhibits several excellent properties such as high aspect ratio, low density at 1.6 g cm^{-3}, very low coefficient of thermal expansion and high tensile strength about 500 MPa [129]. Such unique properties induce numerous applications including antimicrobial films, transparent films, drug delivery, fibres and textiles, supercapacitors and batteries [129]. Cellulose can be classified as microfibrillated celluloses (MFCs), nanofibrillated celluloses (NFCs) and CNCs. However, the hydrophilic nature of nanocellulose can limit its use as the reinforcements in nanocomposite systems where hydrophobic polymers are generally deemed as matrices. This is because blending nanocelluloses with hydrophobic polymers can yield weak interfacial bonding between polymer matrices and nanofillers, resulting in increasing water uptake and filler aggregation with respect to hydrogen bonding [130].

Graphene oxides (GOs) are two-dimensional nanomaterials synthesised from natural graphite. Such nanosheets can be easily prepared from graphite flakes by thermal oxidation, as recommended by Hummers [131]. GOs possess high mechanical properties (e.g. elastic modulus: 0.25 TPa), binding potential, high aspect ratio, high flexibility and superior processibility [132]. Moreover, due to their good hydrophilic characteristic, GOs can be easily exfoliated into monolayer nanosheets stably dispersed in water, primarily due to plenty of hydrophilic oxygenated groups [133]. These functional groups play a significant role in the improvement of GO dispersion in solvents or within polymer matrices through covalent or non-covalent bonding [133]. However, a high tendency of GO agglomeration can limit their use at high nanofiller content levels. Recently, Chang et al. [134] examined the toxicity and biocompatibility of GOs on A549 cells, which suggested that GOs induced the oxidative stress and a slight decrease in cell viability at the high GO content.

1.5 Processing of Bionanocomposites

One of most common methods used for bionanocomposite preparation is solution casting. Its first step involves nanoparticle dispersion in either water or organic solvents by stirring or ultrasonic treatment. Subsequently, polymers and nanoparticles are mixed and stirred together in a solution. The final step is then to evaporate the solvents in order to produce final nanocomposites. Solution casting has been successfully employed in the manufacture of various bionanocomposites such as

PVA/HNT nanocomposites [135] and PVA/MMT nanocomposites [13, 136–139]. The main advantage of this method is that intercalated clay structures in nanocomposites can be formed even for low-polar or non-polar polymers notwithstanding that it has been mostly utilised for water-soluble polymers such as PVA [140]. The reaggregation of nanoparticles such as HNTs in polymer matrices may practically occur during a drying process, particularly after the solvent evaporation. At that stage, the polymer may start to precipitate rather than adhere to HNT surfaces. Hence, the use of coagulation method can improve the HNT dispersion within polymer matrices after solvent evaporation, as illustrated in Fig. 1.11 [141].

The second method is in situ polymerisation, in which layered silicates are swollen in liquid monomers or monomer solutions. The monomers start to migrate to layered silicate galleries so that the polymerisation occurs between intercalated sheets. Heat and radiation can then be used to initiate the reaction by diffusing appropriate initiators or by fixing an organic initiator or a catalyst via the cation exchange prior to the swelling of layered silicates. This technique gives rise to the formation of long polymeric chains within clay galleries [142–144]. Toyota research group was the first to report the synthesis of MMT nanocomposites based on ε-caprolactam monomers at 100 °C when reinforced with Na^+-MMT. Similarly, other research groups also successfully utilised in situ polymerisation for thermoplastics such as PMMA [145], PS [146], polybenzoxale (PBO) [147] and polyolefins like PP and PE [148]. One of major drawbacks from this method is a time-consuming preparation route owing to the fact that the polymerisation can take over a day to complete. The other is the instability of thermodynamic state of exfoliated structures, which may result in the reaggregation of layered silicates [149]. Moreover, in situ polymerisation is also widely used for the preparation of HNT-based nanocomposites. It is proven that the application of such a method can generate covalent bonding between HNTs and polymer matrices [45]. Various HNT-based nanocomposites have been prepared using this method such as the emulsion polymerisation of styrene with HNTs in the presence of sodium dodecyl sulfate (SDS) as an emulsifier (Fig. 1.12) [150]. Direct

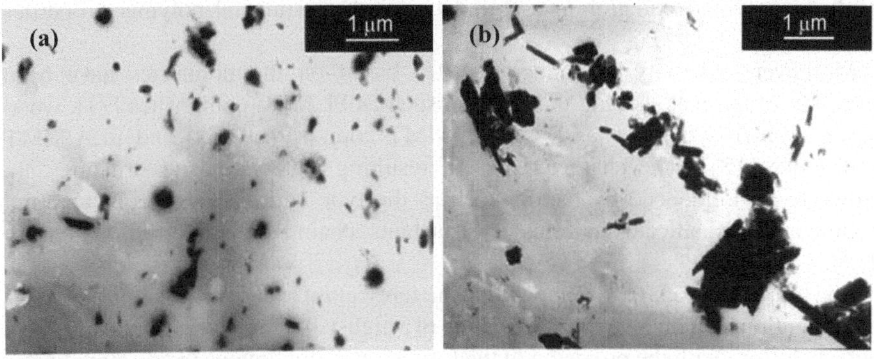

Fig. 1.11 Transmission electron micrographs of PVA/HNT nanocomposites prepared by solution casting process using **a** coagulation and **b** as-cast [141]

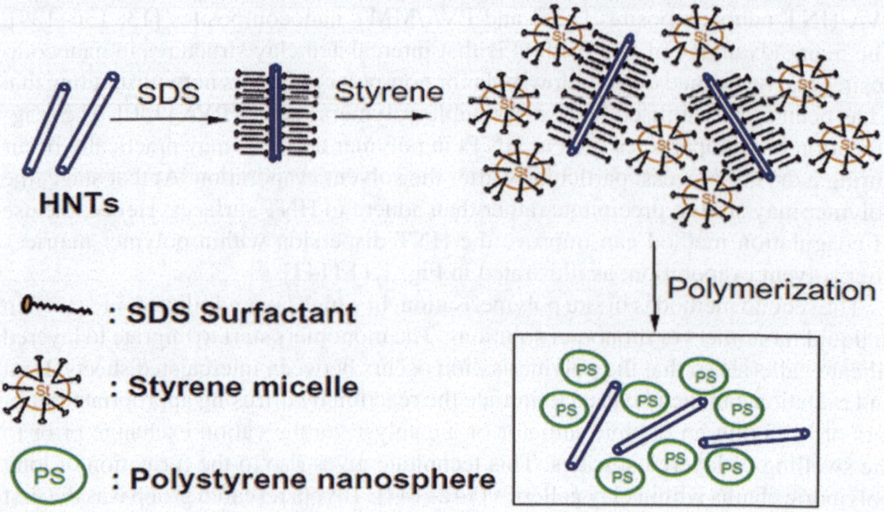

Fig. 1.12 Schematic diagram of in situ polymerisation of styrene in the presence of HNTs [150]

melt intercalation is likely to be the most popular method to disperse nanoparticles into polymer matrices in nanocomposite systems owing to its environmental friendliness and cost-effectiveness in the absence of solvents. This technique involves heating the mixture of polymer pellets and nanoparticles up to a melting temperature T_m of semicrystalline polymers or above T_g of amorphous polymers [151, 152]. Using this process, nanofillers such as HNTs can be mixed with a molten polymer under high shear forces with a variety of mixing devices. At this stage, the molten polymeric chains start to interact with HNTs due to mechanical shearing mechanism, leading to interfacial compatibility with the aid of compression moulding, injection moulding and screw extrusion to manufacture final polymer/HNT nanocomposites. Nonetheless, the combination of high-processing temperature and high shear forces often results in polymer oxidation and degradation with detrimental polymer properties [45].

Moreover, a variety of nanocomposites based on thermoplastics have been successfully fabricated using this method such as PLA/BC composites [31], wood plastic (WP)/BC composites [34] PCL/MMT composites [153] and PLA/MMT composites [154]. The characteristics of resulting nanocomposite structures are known to be influenced by factors such as the type and nature of polymers and modified layered silicates in terms of surfactants, chain length and packing density [155].

The fourth method for bionanocomposite fabrication is electrospinning. As previously mentioned, it involves the application of a high-voltage electrical field into polymeric solutions with the provision of the formation of continuous micro-/nanofibres. These fibres can be used to produce fabric networks with high porosity, small pore sizes and high surface-to-volume ratios. The electrospinning method has been successfully implemented to various nanocomposites systems such as PVA/HNT nanocomposite fibres [156] and PVA/BC composite fibres [157]. In recent years, electrophoretic deposition (EPD) method has also gained a particular attention for the deposition of composite films with HNT inclusions. EPD involves the motion of charged particles in suspension to an electrode leading to the formation of a deposit on the electrode surface induced by uniform electrical-field effect [158]. The main advantage of this method includes its potential to produce uniform films with controlled thickness, as well as for the deposition on substrates with a complex shape and high deposition rate. Such a procedure has been successfully used to fabricate chitosan/HNT nanocomposites [159, 160].

1.6 Bionanocomposite Properties and Characterisation

1.6.1 Morphological Structure

It should be emphasised that mixing a polymer and inorganic nanofillers together does not necessarily produce a nanocomposite. In an immiscible system, poor interfacial bonding between organic and inorganic components yields the separation of nanocomposite constituents into discrete phases in final structures. This separation induces the same properties in so-called nanocomposites as those of microcomposites [142]. In the case of a miscible system, two types of nanocomposites, namely intercalated and exfoliated, (i.e. delaminated) nanocomposites can be achieved, as shown in Fig. 1.13. The determination of final nanocomposite type in terms of intercalated or exfoliated structures depends primarily on synthetic techniques and chemical nature of nanocomposite constituents. In addition, intercalated structures are formed when polymeric chains are penetrated into silicate layers, resulting in a multilayered structure of polymer matrices and inorganic fillers. In this case, polymeric chains expand the interlayer spacing of layered silicates by 20–30 Å. In contrast, exfoliated structures occur when layered silicates are peeled apart and individually separated with an increase in interlayer spacing up to 80–100 Å or above, which is most likely to be associated with the radius of gyration for polymers [142, 161, 162]. Exfoliated structures are generally more favourable because clay dispersibility is significantly promoted as individual particles/fillers leading to greatly enhanced mechanical properties of final nanocomposites. However, it is more difficult to achieve such an ideal structure since the majority of nanocomposites reported in the literatures possess either intercalated or a mix of intercalated and exfoliated structures [136–138, 140]. The reason for this phenomenon is that layered silicates are highly anisotropic with lateral dimensions of 100-1000 nm. As a result, even when separated by the high interlayer spacing, silicate layers cannot be completely and randomly dispersed within polymer matrices [162]. X-ray diffraction (XRD) and transmission electron microscopy (TEM) are the

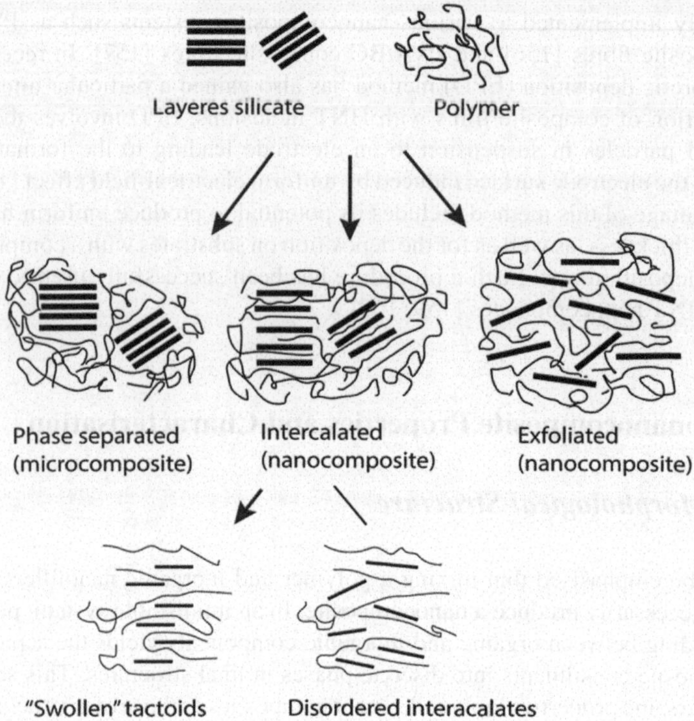

Fig. 1.13 Schematic diagram illustrating different clay structures of polymer/layered silicate nanocomposites [163]

most common techniques used to sophisticatedly characterise final morphological structures of nanocomposites [142, 161]. XRD analysis allows for the determination of the distance between silicate layers in nanocomposite structures according to Bragg's law. For example, in an immiscible polymer/organomodified layered silicate (OMLS) composite system, the characteristics of OMLS basal reflection remain unchanged (Fig. 1.14). However, in a miscible system, the formation of an intercalated structure leads to an increase in interlayer spacing, as evidenced by the shift of XRD peak to lower diffraction angles. On the other hand, the relatively large interlayer spacing for exfoliated structures generally results in a featureless diffraction pattern and disappearance of diffraction peaks in view of XRD spectra. This phenomenon takes place because the large interlayer spacing between silicate layers in such a nanocomposite system tends to result in well-dispersed OMLS with the random orientation [142, 164]. Only XRD analysis would be insufficient for determining final nanocomposite structures. As a result, other techniques such as TEM should be employed in order to offer a visual evidence of morphological structures in nanocomposites. Layered silicates contain heavier elements such as Al, Si and O

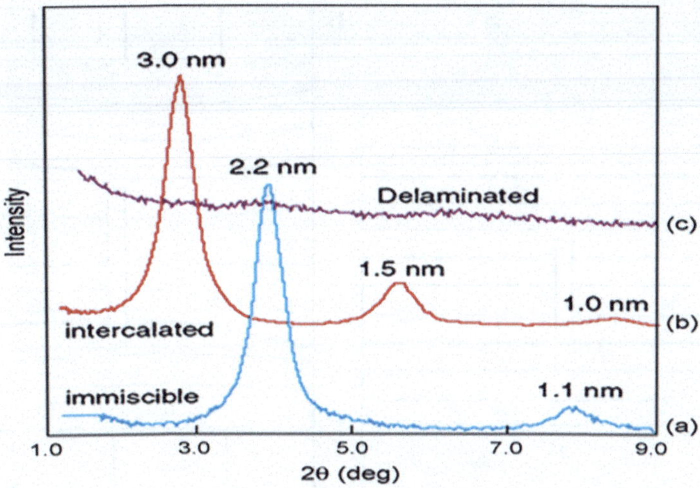

Fig. 1.14 Typical XRD patterns for polymer/layered silicate composites: **a** immiscible structure, **b** intercalated structure and **c** exfoliated structure [142]

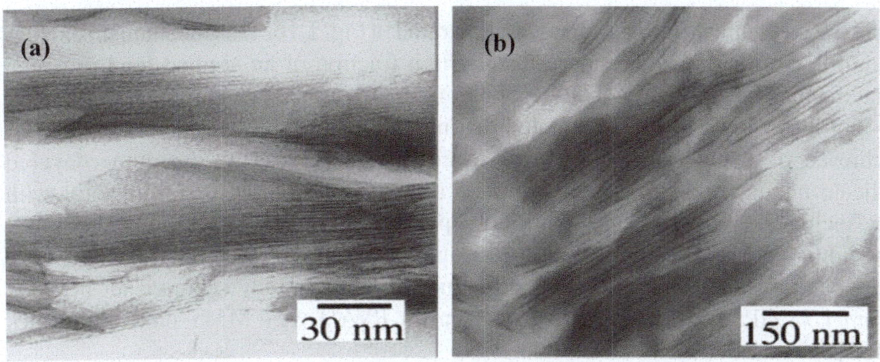

Fig. 1.15 Transmission electron micrographs: **a** intercalated structure and **b** exfoliated structure [165]

when compared with surrounding polymeric chains with only C, H and N. Accordingly, much heavier elements produce the dark appearance for layered silicates in a bright field of TEM images, as illustrated in Fig. 1.15 [142, 161, 164].

1.6.2 Mechanical Properties

The mechanism that explains the reinforcement effect of nanofillers in nanocomposites is the assumption that soft polymer matrices adjacent to much stiffer nanofillers

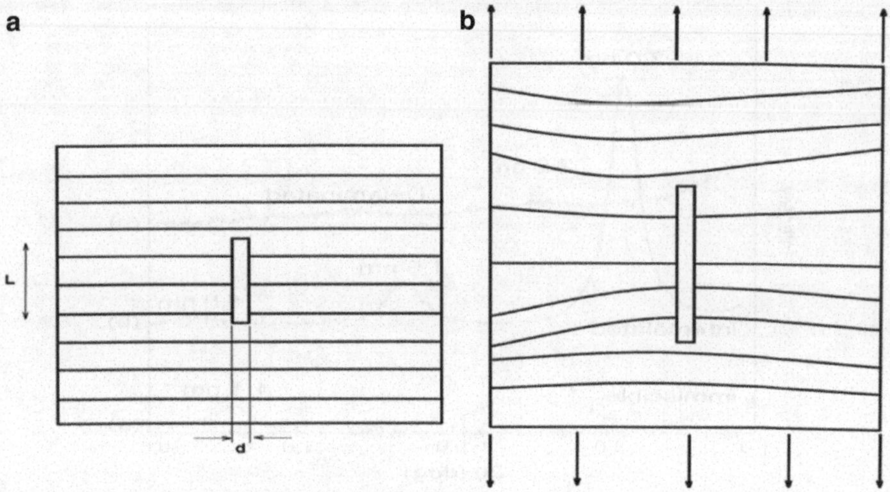

Fig. 1.16 Schematic diagram illustrating the reinforcement mechanism in a composite material: **a** before loading and **b** after loading [142]

become mechanically constrained, as illustrated in Fig. 1.16. This mechanism enables nanofillers to carry a disproportionately high fraction of an applied load provided that the interfacial bonding between polymer matrices and fillers remains sufficient.

Furthermore, it can be postulated that much greater reinforcement effect can be achieved by using fillers with large specific surface areas. In a polymer/layered silicate nanocomposite system, Young's modulus can increase significantly with the inclusion of OMLS into continuous polymer matrices, For example, Young's modulus increased from 1.9 GPa for neat PVA to 3.2 GPa for PVA/5 wt% MMT nanocomposites [166]. In a similar manner, the other study reported an increase from 1.78 GPa for neat PVA to 2.65 GPa for PVA/4 wt% MMT nanocomposites [167].

As can be clearly seen in Table 1.3, Young's modulus of nanocomposites is a function of clay content. However, any further increase in the clay content appears to make a declining effect on Young's modulus. Based on XRD and TEM observations, this trend is attributed to the change of clay dispersion states from full exfoliation, a mix of exfoliation and intercalation to clay agglomeration at the clay content of 10 wt% or above [142, 143].

Apart from the change in Young's modulus, the incorporation of organoclays into polymer matrices also causes the increase in tensile strength of nanocomposites. Wang et al. [166] reported that the tensile strength of PVA nanocomposites was increased by 52% at the MMT loading of 3 wt% as opposed to that of neat polymer. Similarly, another investigation [168] reported that the tensile strength increased from 50 MPa for neat PVA to 95.9 MPa for PVA-based nanocomposites at the clay loading of 17.3 wt%. Nonetheless, it should be clearly stated that the reduction in tensile strength for nanocomposites has also been noted, Table 1.3. This result can be

Table 1.3 Processing and mechanical properties of different polymer/layered silicate nanocomposites

Polymer	Process	Filler type and content (wt%)	Young's modulus (MPa)	Tensile strength (MPa)	References
PVA	Solvent intercalation	MMT (0)	1900	65.1	[166]
		MMT (3)	3100	98.9	
		MMT (5)	3200	91.6	
PVA	Solvent intercalation	Na^+ MMT (0)	1470	52.25	[167]
		Na^+ MMT (4)	2160	63.36	
		Na^+ MMT (8)	2690	54.95	
PVA/starch	Melt intercalation	Na^+ MMT (0)	9.65	11	[169]
		Na^+ MMT (5)	23.76	13.3	
		Na^+ MMT (10)	30.58	14.7	
		Na^+ MMT (15)	39.17	16.17	
		Na^+ MMT (20)	47.67	14.47	
PVA/PVP	Solvent intercalation	Na^+ MMT (0)	1780	42.4	[167]
		Na^+ MMT (4)	2650	48.9	
		Na^+ MMT (8)	4580	52.3	
GS/PVA	Solvent intercalation	MMT (0)	197.8	38.5	[170]
		MMT (5)	265	48	
		MMT (10)	315	57.5	
		MMT (15)	331.2	60.2	
		MMT (20)	225.5	42.3	
PVA	Solvent intercalation	MMT (0)	2064	50	[168]
		MMT (1.84)	3640	99.6	
		MMT (5.44)	2966	73.3	
		MMT (17.3)	5055	95.9	
GS/PVA	Solvent intercalation	HNT (0)	990	40	[15]
		HNT (0.25)	1300	46	
		HNT (0.5)	1560	51	
		HNT (3)	1700	32	
		HNT (5)	1860	29	
PVA- MA		HNT (0)	1023	46.6	[16]
		HNT (5)	1105	42.1	
		HNT (10)	1185	38.8	
		HNT (20)	1244	36.1	

Note: *GS* chitosan and *MA* malonic acid

ascribed to clay agglomeration resulting from increasing the clay content generally over 5 wt% associated with non-uniform clay dispersion. Those clay aggregates act as typical defects of stress-concentration sites for crack initiation [142].

The molecular weight of polymer matrices also has an impact on reinforcement efficiency and tensile strength of nanocomposites. It has been reported that Young's modulus and yield strength of nanocomposites are substantially enhanced with increasing the molecular weight of polymer matrices [142]. This finding suggests that increasing the molecular weight of polymer matrices induces higher shear stress to break up large clay particles into exfoliated platelets. The higher degree of exfoliation inevitably results in effective stress transfer from polymer matrices to nanofillers with their stronger interfacial bonding.

Other factors to play a key role in enhancing Young's modulus and tensile strength of nanocomposites are organic modification of layer silicates and the use of compatibilisers to polymer matrices. Hotta and Paul [171] indicated that increasing the number of alkali tails for organoclays from one tail to two tails enhanced Young's modulus and tensile strength of nanocomposites. Pavlidou and Papaspyrides [142] also demonstrated that increasing the amount of CAB-MA compatibiliser in cellulose acetate butyrate (CAB)/CAB-MA-based nanocomposites from 0 to 5 wt% enhanced Young's modulus from 3.6 to 3.7 GPa [142].

In polymer/HNT nanocomposite systems, the improvements of elastic modulus and tensile strength depend primarily on the HNT loading, HNT dispersion and interfacial adhesion between HNTs and polymer matrices. As indicated in Table 1.3, increasing the HNT loading generally results in higher reinforcement efficiency, as well as greater tensile strength of nanocomposites. However, these trends are not always observed due to the influence of HNT dispersion within polymer matrices. Although HNTs are more easily dispersed when compared to natural silica and CNTs, their dispersion can still be challenging, particularly for non-polar polymers [172, 173]. Qiu et al. [16] reported that the tensile strength of PVA/HNT nanocomposites was decreased by 14.2% when their loading increased from 5 to 10 wt% due to the HNT agglomeration at 10 wt%. Better dispersion using surface pre-treatment, including the grafting of silanes or polymers, can yield the improvement of interfacial bonding (i.e. hydrogen bonding) between HNTs and polymer matrices. It was shown that the modulus of PVA/HNT nanocomposites was increased by 21% when using modified HNTs as opposed to unmodified HNTs [14].

1.6.3 Nanomechanical Properties of Nanocomposites

Nanocomposites represent a homogenous material system at the macroscaled level despite their heterogeneity on the micro- or nanoscale. In general, polymers react with nanoparticles during a mixing process to yield multiple phases including partially or fully reacted composites, unreacted components of raw materials and interphase zones, which have different mechanical properties and chemical properties [174]. The properties and amount of each phase relative to entire nanocomposites strongly affect the properties of bulk materials [175]. Therefore, only using conventional mechanical testing is insufficient to evaluate the real reinforcement effect of nanoparticles on polymer matrices, as well as detect localised reinforcement and deformation.

Additionally, even the results of conventional mechanical testing is directly based on a fundamental concept that the chain mobility of soft matrices is constrained by much stiffer nanoparticles [25]. As a consequence, an effective load transfer occurs from matrices to fillers such as nanoparticles to carry a disproportionally high fraction of applied loads, thus leading to an increase in load resistance [25]. This phenomenon is well known to be associated with the level of interfacial bonding between nanoparticles and polymer matrices in terms of interphase existence, dimensions, structures and compositions [176]. Hence, the crack initiation or propagation may take place in a nanocomposite system, resulting from the lack of effective load transfer owing to weak interfacial bonding.

Interfaces are described as a material boundary between two or more phases with distinct chemical/physical properties and morphological structures. Furthermore, a material volume influenced by the interfacial interaction can be named 'interphase' [177]. Interphase regions start from the interfacial boundary of nanofillers with different properties from those of bulk nanofillers and end where they are in connection with polymer matrices whose properties also vary from bulk matrices [178]. The material performance of nanocomposites is primarily impacted by the interphase regions where structural and chemical changes such as cross-linking density and crystalline phases result in major alteration to composite bulk properties [177]. It is also worth mentioning that the change to the mobility of polymeric chains plays an important role in mechanical and dielectric properties of nanocomposites [175]. Due to large interfacial-area-to-volume ratios in nanocomposites, interfacial regions consist of a significant portion of bulk nanocomposites. For instance, with the addition of 5 vol% monodispersed spherical nanoparticles (particle diameter: 10 nm and interphase thickness: 0.5 nm), the volume fraction of interphase can be as high as 25 vol% [175]. More impressively, when particle diameter is reduced to less than 5 nm, the volume fraction of interphases was increased by over 50 vol% as compared to that of particles [175]. The characterisation of existing interphase and its associated properties is generally difficult to undertake as the interphase for nanocomposites is generally on a nanoscaled level, and thus researchers have had to make most experimental efforts in an uncrosslinked state as the indirect evidence [179]. For example, chain mobility near interphase regions can be less than those of polymer matrices in a nanocomposite system [179]. As a result, Litvinov and Steeman [180] employed proton, low-resolution T2 nuclear magnetic resonance (NMR) relaxation technique to detect existing interphases between ethylene propylene dine monomers (EPDMs) and carbon black. It was indicative of a significant difference in the chain mobility of EPDMs near carbon black surfaces where the generated layer sizes of immobilised EPDMs were estimated in range of 1–2 of the unit diameter of monomers [180]. Pompe and Mäder [181] identified the interphase according to differential scanning calorimetry (DSC) in PP/glass fibre composites, which, however, was limited to semicrystalline polymers as matrices in composite materials at only high glass fibre contents. Brown et al. [182] studied a relationship between nanoparticle diameter and interphase thickness based on molecular dynamic (MD) simulations, which implied that the interphase thickness appeared to be relatively insensitive to nanoparticle diameters and contents. Li et al. [183] reported that the volume fraction of interphases could be size independent by using a modified hierarchical multi-interphase model (MHMM). However, the interphase thickness might be influenced by the reinforcing efficiency of nanoparticles when their lengths were over 40 nm. On the contrary, Gu et al. [184] and other co-workers [185, 186] inferred that the interphase thickness could be non-constant in nanocomposites systems. So far, nanointerphase properties and features have not been explicitly quantified in a systematic manner by means of direct topography of nanomechanical characterisation. As a matter of fact, nanomechanical techniques using tip–sample interactions such as atomic force microscopy (AFM) [187], nanoindentation and nanoscratch tests [188, 189] are vital as effective and relatively straightforward approaches to determine nanomechanical properties of interphase. In addition, interphase dimensions and sizes can be clearly specified due to their distinct mechanical properties from those of bulk materials. The interphase width of glass fibres coated with coupling agents was previously determined to be around several microns [188, 190]. Nonetheless, a quantitative analysis of interphase dimensions associated with nanomechanical properties still undergoes limited lateral resolutions and positioning capability of indenter probe used in nanoindentation and nanoscratch techniques. In particular, interphase regions of thermosetting polymer/glass fibre composites are much thinner than those of individual fibre and matrix components [191]. Plastic deformation usually takes place under a high fraction of applied load, thus resulting in an increase in the minimum allowable distance between tow indentation spots, as well as the reduction of lateral resolutions of such tests [190, 192]. The peak force quantitative nanomechanical mapping (PFQNM) becomes a relatively new and powerful technique to quantitatively measure nanomechanical properties of materials such as the stiffness and adhesion of nanocomposites along with corresponding acquired dimensions [193, 194]. The use of PFQNM greatly supports the measurement of material elastic properties based on tip–sample force curves and the acquisition of topographic images simultaneously. Moreover, other critical properties, consisting of tip–surface adhesion and surface deformation, can also be obtained by overcoming the difficulty associated with lateral forces. Such a technique is believed not only to sophisticatedly distinguish between nanofillers, nanointerphases and polymer matrices, but also to accurately quantify dimensions and nanomechanical properties of interphase.

Nanomechcanical properties of PVA and PVA nanocomposites are completely different from those of their bulk material counterparts. It is worth noting that average elastic modulus of bulk PVA films is approximately 2.064 GPa at a macroscopic level [168], which is far less in magnitude when compared with local nanophase such as 9.9 GPa for PVA/10 wt% poly(acrylic acid) (PAA) blends [195]. Moreovere, in case of PVA-PAA-based nanocomposites reinforced with 10 wt% CNCs [195], their average interphase elastic modulus was found to vary from 12.8 GPa at the interface of CNCs to 9.9 GPa in PVA-PPA matrices. On the contrary, PVA nanocomposites reinforced with 10 wt% CNCs possessed the highest elastic modulus only 1.9 GPa in their bulk properties [196]. Such a modulus-variation phenomenon between nanomechanical properties and bulk properties is not limited to PVA alone, but can be applied to different polymers or composites. For instance, the storage modulus of epoxy was measured to be 17 GPa via nanomechanical measurement as opposed to 3–4 GPa for their bulk composites determined by tensile tests [184]. Moreover, elastic modulus of bulk poly(ether sulfone) (PESU) membrane substrates was found to be only 151.7 ± 7.9 MPa [197] when compared to 3.2 ± 0.3 GPa for PESU films at a nanoscaled level [198].

1.6.4 Thermal Properties

The thermal properties of polymeric materials can be quantified by several techniques. Among these, thermogravimetric analysis (TGA) and DSC are the most popular characterisation techniques. TGA is used to measure the weight loss resulting from the formation of volatile products or chemical reactions as a function of temperature and/or time, while DSC is utilised to measure T_g, crystallisation temperature (T_c) and T_m of polymeric materials. In a polymer/clay nanocomposite system, the inclusion of layered silicates within polymer matrices can improve the thermal stability by acting as an insulator and mass transport barrier to volatile products that are generated through the degradation process [25]. For instance, the decomposition temperature of PVA/5wt% MMT nanocomposites has increased to 354 °C as opposed to 337 °C for pure PVA [13]. Moreover, the T_g of PVA was also increased from 34 to 60 °C when MMT content increased up to 10 wt%. Such a finding was related to uniform dispersion of MMTs, which dissipated the applied heat through the polymer, thereby leading to the enhanced heat resistance for final nanocomposites [13]. In addition, Mallakpour et al. [139] reported that the degradation of PVA was promoted with the inclusion of fluorohectorite or MMTs. The char yield of PVA at 800 °C was also increased from 6% to 19% in case of PVA/5 wt% MMT nanocomposites, implying the improvement of thermal stability at the higher temperature [139]. Another study showed that the thermal decomposition temperature of PVA/OMLS nanocomposites shifted to a marginally higher temperature compared with that of pure PVA [199]. This phenomenon was explained according to Stawhecker and Manias [140] that PVA could supply oxygen by itself to initiate the thermal decomposition. Conversely, the reduction in the thermal stability of nanocomposites has also been noted as stack-layered silicates can accumulate the heat at the early stages of decomposition [142]. Alkali ammonium cations existing in OMLS may experience subsequent decomposition by Hofmann elimination leading to the acceleration of polymer degradation [4, 142]. Katti et al. [200] indicated that the addition of MMTs and hydroxyapatite (HAP) to chitosan improved the thermal stability of such new nanocomposites with even higher onset temperature and the percentage residue left at 500 °C when compared to chitosan/MMT nanocomposites.

In case of polymer/HNT nanocomposite system, thermal properties may be significantly improved compared to polymer matrices due to the following reasons: (i) the thermal stability of HNTs is much higher than that of polymer matrices with initial degradation typically taking place at ~400 °C; (ii) excellent HNT dispersion in polymer matrices leads to better barrier effect towards mass and heat transport and (iii) polymer chains and volatile products can enter HNT lumens leading to a delay of mass transport with the further improvement of thermal stability. Swapna et al. [201] found that the inclusion of HNTs improved thermal stability of PVA due to the heat-barrier effect of inorganic fillers. However, T_g and the degree of crystallinity were decreased with the inclusion of HNTs, possibly relating to the plasticisation effect of hydrophilic HNT fillers [201]. Qiu et al. [16] stated that decomposition temperatures at the weight loss of 30 and 50% appeared to be increased by 3–5 °C for PVA/HNT nanocomposite films when compared with neat PVA despite no identified change of T_g. This phenomenon mainly depends on uniform dispersion of HNTs within PVA matrices for better thermal stability. Similarly, other studies have also indicated a critical role of HNTs in improving thermal properties of various biopolymers such as PVA [16], PEG [202] and starch [203].

1.7 Summary

The preparation and properties of different bionanocomposites reinforced with HNTs, OMLS and BCs have been explicitly discussed. Many advantages of bionanocomposites are summarised as follows:

- Improved mechanical properties of bionanocomposites in solid and melt states as opposed to those based on conventional composites.
- Significant increase in the thermal stability of bionanocomposites with the addition of nanofillers to work as heat barriers in the formation of char after thermal degradation.
- Improved biodegradability of some biopolymers with the inclusion of nanofillers.

The above-mentioned property improvements generally occur at a much lower loading of nanofillers than those of conventional fillers in composite systems. Therefore, bionanocomposites not only offer far more lightweight products when compared with conventional composites, but also make a superior contribution to widespread applications owing to their excellent material characteristics. Despite numerous studies and research conducted in the past decades, the commercialisation of bionanocomposites still undergoes enormous challenges encountered in material development and innovation in order to meet the applications required by the end users.

References

1. Geyer R, Jambeck JR, Law KL (2017) Production, use, and fate of all plastics ever made. Sci Adv 3(7):1–6
2. Reddy MM, Misra M, Mohanty AK (2012) Bio-based materials in the new bio-economy. Chem Eng Prog 108(5):37–42
3. Averous L, Boquillon N (2004) Biocomposites based on plasticized starch: thermal and mechanical behaviours. Carbohydr Polym 56(2):111–122
4. Ray SS, Bousmina M (2005) Biodegradable polymers and their layered silicate nanocomposites: in greening the 21st century materials world. Prog Mater Sci 50(8):962–1079
5. Robyt JF (2008) Starch: structure, properties, chemistry, and enzymology. In: Fraser-Reid BO, Tatsuta K, Thiem J (eds) Glycoscience: chemistry and chemical biology. Springer, Berlin, pp 1437–1472
6. Eagan JM, Xu J, Di Girolamo R, Thurber CM, Macosko CW, LaPointe AM, Bates FS, Coates GW (2017) Combining polyethylene and polypropylene: enhanced performance with PE/iPP multiblock polymers. Science 355(6327):814–816
7. Xanthos D, Walker TR (2017) International policies to reduce plastic marine pollution from single-use plastics (plastic bags and microbeads): a review. Mar Pollut Bull 118(1–2):17–26
8. Avérous L (2004) Biodegradable multiphase systems based on plasticized starch: a review. J Macromol Sci Part C: Polym Rev 44(3):231–274
9. Huang D, Wang A (2013) Non-covalently functionalized multiwalled carbon nanotubes by chitosan and their synergistic reinforcing effects in PVA films. RSC Adv 3(4):1210–1216
10. Pan Y-S, Xiong D-S, Ma R-Y (2007) A study on the friction properties of poly (vinyl alcohol) hydrogel as articular cartilage against titanium alloy. Wear 262(7–8):1021–1025
11. Espinosa HD, Rim JE, Barthelat F, Buehler MJ (2009) Merger of structure and material in nacre and bone–Perspectives on de novo biomimetic materials. Prog Mater Sci 54(8):1059–1100
12. Sapalidis AA, Katsaros FK, Kanellopoulosm NK (2011) PVA/montmorillonite nanocomposites: development and properties. In: Cuppoletti J (ed) Nanocomposites and polymers with analytical methods. InTech, Rijeka, pp 29–50
13. Jose T, George SC, Maria HJ, Wilson R, Thomas S (2014) Effect of bentonite clay on the mechanical, thermal, and pervaporation performance of the poly (vinyl alcohol) nanocomposite membranes. Ind Eng Chem Res 53(43):16820–16831
14. Fujii K, Nakagaito AN, Takagi H, Yonekura D (2014) Sulfuric acid treatment of halloysite nanoclay to improve the mechanical properties of PVA/halloysite transparent composite films. Compos Interfaces 21(4):319–327
15. Khoo W, Ismail H, Ariffin A (2013) Tensile, swelling, and oxidative degradation properties of crosslinked polyvinyl alcohol/chitosan/halloysite nanotube composites. Int J Polym Mater Polym Biomater 62(7):390–396
16. Qiu K, Netravali AN (2013) Halloysite nanotube reinforced biodegradable nanocomposites using noncrosslinked and malonic acid crosslinked polyvinyl alcohol. Polym Compos 34(5):799–809
17. Du F-P, Ye EZ, Yang W, Shen TH, Tang CY, Xie XL, Zhou XP, Law WC (2015) Electroactive shape memory polymer based on optimized multi-walled carbon nanotubes/polyvinyl alcohol nanocomposites. Compos Part B Eng 68:170–175
18. Minus ML, Chae HG, Kumar S (2009) Interfacial crystallization in gel-spun poly (vinyl alcohol)/single-wall carbon nanotube composite fibers. Macromol Chem Phy 210(21):1799–1808
19. Thayumanavan N, Tambe P, Joshi G (2015) Effect of surfactant and sodium alginate modification of graphene on the mechanical and thermal properties of polyvinyl alcohol (PVA) nanocomposites. Cellul Chem Technol 49:69–80
20. Li YQ, Yu T, Yang TY, Zheng LX, Liao K (2012) Bio-inspired nacre-like composite films based on graphene with superior mechanical, electrical, and biocompatible properties. Adv Mater 24(25):3426–3431

21. Cheng Q, Wang S, Tong Z (2015) Poly (vinyl alcohol) cellulose nanocomposites. In: Kar KK, Pandey, JK, Rana SK (eds) in Handbook of polymer nanocomposites. Processing, performance and application, volume C: polymer nanocomposites of cellulose nanoparticles. Springer, Berlin, pp 433–447
22. Leitão AF, Silva J, Dourado F, Gama M (2013) Production and characterization of a new bacterial cellulose/poly (vinyl alcohol) nanocomposite. Materials 6(5):1956–1966
23. Ren W, Wu R, Guo P, Zhu J, Li H, Xu S, Wang J (2015) Preparation and characterization of covalently bonded PVA/Laponite/HAPI nanocomposite multilayer freestanding films by layer-by-layer assembly. J Polym Sci, Part B: Polym Phys 53(8):545–551
24. Morimune S, Kotera M, Nishino T, Goto K, Hata K (2011) Poly (vinyl alcohol) nanocomposites with nanodiamond. Macromolecules 44(11):4415–4421
25. Mousa M, Dong Y (2020) Polymeric materials as bionanocomposites. In: Zai KM, Jabeen F, Anjum MN, Ikram S (eds) Bionanocomposites: green synthesis and applications. Elsevier, Kidlington, pp 335–365
26. Putz KW, Compton OC, Palmeri MJ, Nguyen ST, Brinson LC (2010) High-nanofiller-content graphene oxide–polymer nanocomposites via vacuum-assisted self-assembly. Adv Funct Mater 20(19):3322–3329
27. Liao P, Ismael ZM, Zhang W, Yuan S, Tong M, Wang K, Bao J (2012) Adsorption of dyes from aqueous solutions by microwave modified bamboo charcoal. Chem Eng J 195:339–346
28. Singh K, Singh RS, Rai BN, Upadhyay SN (2010) Biofiltration of toluene using wood charcoal as the biofilter media. Bioresource Technol 101(11):3947–3951
29. Asada T, Ohkubo T, Kawata K, Oikawa K (2006) Ammonia adsorption on bamboo charcoal with acid treatment. J Health Sci 52(5):585–589
30. Lou C-W, Lin CW, Lei CH, Su KH, Hsu CH, Liu ZH, Lin JH (2007) PET/PP blend with bamboo charcoal to produce functional composites. J Mater Process Technol 192–193:428–433
31. Ho M-P, Lau K-T, Wang H, Hui D (2015) Improvement on the properties of polylactic acid (PLA) using bamboo charcoal particles. Compos Part B Eng 81:14–25
32. Yeh JT, Hsiung HH, Wei W, Zhu P, Chen KN, Jiang T (2009) Negative air ion releasing properties of tourmaline/bamboo charcoal compounds containing ethylene propylene diene terpolymer/polypropylene composites. J Appl Polym Sci 113(2):1097–1110
33. Zhao R-S, Wang X, Yuan JP, Lin JM (2008) Investigation of feasibility of bamboo charcoal as solid-phase extraction adsorbent for the enrichment and determination of four phthalate esters in environmental water samples. J Chromatogr A 1183(1–2):15–20
34. Li X, Lin Z, Huang L, Tan S, Cai X (2014) The utilization of bamboo charcoal enhances wood plastic composites with excellent mechanical and thermal properties. Mater Design 53:419–424
35. Nitayaphat W, Jiratumnukul N, Charuchinda S, Kittinaovarat S (2009) Mechanical properties of chitosan/bamboo charcoal composite films made with normal and surface oxidized charcoal. Carbohydr Polym 78(3):444–448
36. Yang F-C, Wu KH, Lin WP, Hu MK (2009) Preparation and antibacterial efficacy of bamboo charcoal/polyoxometalate biological protective material. Microporous Mesoporous Mater 118(1–3):467–472
37. Kamada K (2008) Study on healthy housing using recycled organic industrial waste first report: overview of trends in the development of the latest technology and new materials in Japan. No. 32, Hokkaido Bunkyo University, pp 1–17
38. Wu J, Wei Y, Lin J, Lin S (2003) Preparation of a starch-graft-acrylamide/kaolinite superabsorbent composite and the influence of the hydrophilic group on its water absorbency. Polym Int 52(12):1909–1912
39. Lin J, Wu J, Yang Z, Pu M (2001) Synthesis and properties of poly (acrylic acid)/mica superabsorbent nanocomposite. Macromol Rapid Commun 22(6):422–424
40. Zhang F, Guo Z, Gao H, Li Y, Ren L, Shi L, Wang L (2005) Synthesis and properties of sepiolite/poly (acrylic acid-co-acrylamide) nanocomposites. Polym Bull 55(6):419–428

41. Ferfera-Harrar H, Aiouaz N, Dairi N, Hadj-Hamou AS (2014) Preparation of chitosan-g-poly (acrylamide)/montmorillonite superabsorbent polymer composites: studies on swelling, thermal, and antibacterial properties. J Appl Polym Sci 131(1):39747
42. Filippi S, Paci M, Polacco G, Dintcheva NT, Magagnini P (2011) On the interlayer spacing collapse of Cloisite® 30B organoclay. Polym Degrad Stab 96(5):823–832
43. Tavares LL, Almeida CB, Caruso ÍP, Cornélio ML, Lopes Filho JF (2012) Effect of modified clays on the structure and functional properties of biofilms produced with zein. Food Sci Technol 32(2):314–322
44. Rhim J-W, Ng PK (2007) Natural biopolymer-based nanocomposite films for packaging applications. Crit Rev Food Sci Nutr 47(4):411–433
45. Liu M, Jia Z, Jia D, Zhou C (2014) Recent advance in research on halloysite nanotubes-polymer nanocomposite. Prog Polym Sci 39(8):1498–1525
46. Jem KJ, van der Pol JF, de Vos S (2010) Microbial lactic acid, its polymer poly (lactic acid), and their industrial applications. In: Chen GG-Q (ed) Plastics from bacteria: natural functions and applications. Springer, Berlin, pp 323–346
47. Martin O, Averous L (2001) Poly (lactic acid): plasticization and properties of biodegradable multiphase systems. Polymer 42(14):6209–6219
48. Ho KLG, Pometto AL, Gadea-Rivas A, Briceño JA, Rojas A (1999) Degradation of polylactic acid (PLA) plastic in Costa Rican soil and Iowa state university compost rows. J Environ Polym Degrad 7(4):173–177
49. Hakkarainen M (2002) Aliphatic polyesters: abiotic and biotic degradation and degradation products. In: Albertsson A-C (ed) Degradable aliphatic polyesters. Springer, Berlin, pp 113–138
50. Kawaguchi Y, Doi Y (1992) Kinetics and mechanism of synthesis and degradation of poly (3-hydroxybutyrate) in *Alcaligenes eutrophus*. Macromolecules 25(9):2324–2329
51. Lenz RW, Marchessault RH (2005) Bacterial polyesters: biosynthesis, biodegradable plastics and biotechnology. Biomacromol 6(1):1–8
52. Bastioli C, Cerutti A, Guanella I, Romano G, Tosin M (1995) Physical state and biodegradation behavior of starch-polycaprolactone systems. J Environ Polym Degrad 3(2):81–95
53. Averous L, Moro L, Dole P, Fringant C (2000) Properties of thermoplastic blends: starch–polycaprolactone. Polymer 41(11):4157–4167
54. Koenig M, Huang S (1994) Evaluation of crosslinked poly (caprolactone) as a biodegradable, hydrophobic coating. Polym Degrad Stab 45(1):139–144
55. Goldberg D (1995) A review of the biodegradability and utility of poly (caprolactone). J Enviro Polym Degrad 3(2):61–67
56. Bastioli C (1998) Biodegradable materials—present situation and future perspectives. Macromol Symp 135(1):193–204
57. Okada M (2002) Chemical syntheses of biodegradable polymers. Prog Polym Sci 27(1):87–133
58. Gross RA, Kalra B (2002) Biodegradable polymers for the environment. Science 297(5582):803–807
59. Wolk SK, Swift G, Paik YH, Yocom KM, Smith RL, Simon ES (1994) One-and two-dimensional nuclear magnetic resonance characterization of poly (aspartic acid) prepared by thermal polymerization of L-aspartic acid. Macromolecules 27(26):7613–7620
60. Hallensleben ML (1992) Polyvinyl compounds, others. In: Elvers B, Hawkins S, Schulz G (eds) Ullmann's encyclopedia of industrial chemistry, vol A21. VCH Publishers, New York, pp 743–758
61. Baker MI, Walsh SP, Schwartz Z, Boyan BD (2012) A review of polyvinyl alcohol and its uses in cartilage and orthopedic applications. J Biomed Mater Res Part B Appl Biomater 100(5):1451–1457
62. Finch CA (1973) Polyvinyl alcohol, properties and applications. Wiley, New York, pp 340–389
63. Oviedo IR, Méndez NN, Gómez MG, Rodríguez HC, Martínez AR (2008) Design of a physical and nontoxic crosslinked poly (vinyl alcohol) hydrogel. Int J Polym Mater 57(12):1095–1103

64. Briscoe B, Luckham P, Zhu S (2000) The effects of hydrogen bonding upon the viscosity of aqueous poly (vinyl alcohol) solutions. Polymer 41(10):3851–3860
65. Chang I-S, Kim C-I, Nam BU (2005) The influence of poly-vinyl-alcohol (PVA) characteristics on the physical stability of encapsulated immobilization media for advanced wastewater treatment. Process Biochem 40(9):3050–3054
66. Hara C, Matsuo M (1995) Phase separation in aqueous poly (vinyl alcohol) solution. Polymer 36(3):603–609
67. Liu M, Cheng R, Qian R (1995) Effect of solution concentration on the gelation of aqueous polyvinyl alcohol solution. J Polym Sci, Part B: Polym Phys 33(12):1731–1735
68. Nozakura SI, Kida S (1974) Polymerization of vinyloxyaluminum and formation of syndiotactic poly (vinyl alcohol). J Polym Sci, Part A: Polym Chem 12(10):2337–2348
69. Napper D (1970) Flocculation studies of sterically stabilized dispersions. J Colloid Interf Sci 32(1):106–114
70. Maeda H, Kawai T, Sekii S (1959) Intra-and intermolecular hydrogen bonds in polyvinyl alcohol solutions. J Polym Sci, Part A: Polym Chem 35(128):288–292
71. Miranda-Trevino JC, Coles CA (2003) Kaolinite properties, structure and influence of metal retention on pH. Appl Clay Sci 23(1–4):133–139
72. Chiu C-W, Huang TK, Wang YC, Alamani BG, Lin JJ (2014) Intercalation strategies in clay/polymer hybrids. Progress Polym Sci 39(3):443–485
73. Beyer G (2002) Nanocomposites: a new class of flame retardants for polymers. Plast Addit Compound 4(10):22–28
74. Solomon MJ, Almusallam AS, Seefeldt KF, Somwangthanaroj A, Varadan P (2001) Rheology of polypropylene/clay hybrid materials. Macromolecules 34(6):1864–1872
75. Nuruzzaman M, Rahman MM, Liu Y, Naidu R (2016) Nanoencapsulation, nano-guard for pesticides: a new window for safe application. J Agric Food Chem 64(7):1447–1483
76. Alexandre M, Dubois P (2000) Polymer-layered silicate nanocomposites: preparation, properties and uses of a new class of materials. Mater Sci Eng R Rep 28(1–2):1–63
77. Francis CW (1973) Adsorption of polyvinylpyrrolidone on reference clay minerals. Soil Sci 115(1):40–54
78. Aranda P, Ruiz-Hitzky E (1992) Poly (ethylene oxide)-silicate intercalation materials. Chem Mater 4(6):1395–1403
79. Fischer H (2003) Polymer nanocomposites: from fundamental research to specific applications. Mater Sci Eng C 23(6–8):763–772
80. Zanetti M, Lomakin S, Camino G (2000) Polymer layered silicate nanocomposites. Macromol Mater Eng 279(1):1–9
81. Chin I-J, Thurn-Albrecht T, Kim H-C, Russell TP, Wang J (2001) On exfoliation of montmorillonite in epoxy. Polymer 42(13):5947–5952
82. Xu L, Reeder S, Thopasridharan M, Ren J, Shipp DA, Krishnamoorti R (2005) Structure and melt rheology of polystyrene-based layered silicate nanocomposites. Nanotechnology 16(7):S514–S521
83. Wang KH, Choi MH, Koo CM, Choi YS, Chung IJ (2001) Synthesis and characterization of maleated polyethylene/clay nanocomposites. Polymer 42(24):9819–9826
84. Kerr PF (1952) Formation and occurrence of clay minerals. Clays Clay Miner 1(1):19–32
85. Rawtani D, Agrawal Y (2012) Multifarious applications of halloysite nanotubes: a review. Rev Adv Mater Sci 30(3):282–295
86. Guo B, Zou Q, Lei Y, Jia D (2009) Structure and performance of polyamide 6/halloysite nanotubes nanocomposites. Polym J 41(10):835–842
87. Du M, Guo B, Jia D (2010) Newly emerging applications of halloysite nanotubes: a review. Polym Int 59(5):574–582
88. Lecouvet B, Horion J, D'haese C, Bailly C, Nysten B (2013) Elastic modulus of halloysite nanotubes. Nanotechnology 24(10):105704
89. Guimaraes L, Enyashin AN, Seifert G, Duarte HA (2010) Structural, electronic, and mechanical properties of single-walled halloysite nanotube models. J Phys Chem C114 (26):11358–11363

90. Liu M, Guo B, Du M, Cai X, Jia D (2007) Properties of halloysite nanotube–epoxy resin hybrids and the interfacial reactions in the systems. Nanotechnology 18(45):455703
91. Pasbakhsh P, Churchman GJ, Keeling JL (2013) Characterisation of properties of various halloysites relevant to their use as nanotubes and microfibre fillers. Appl Clay Sci 74:47–57
92. Du M, Guo B, Wan J, Zou Q, Jia D (2010) Effects of halloysite nanotubes on kinetics and activation energy of non-isothermal crystallization of polypropylene. J Polym Res 17(1):109
93. Ismail H, Salleh S, Ahmad Z (2013) Properties of halloysite nanotubes-filled natural rubber prepared using different mixing methods. Mater Design 50:790–797
94. Wei W, Minullina R, Abdullayev E, Fakhrullin R, Mills D, Lvov Y (2014) Enhanced efficiency of antiseptics with sustained release from clay nanotubes. RSC Adv 4(1):488–494
95. Yuan P, Tan D, Aannabi-Bergaya F, Yan W, Fan M, Liu D, He H (2012) Changes in structure, morphology, porosity, and surface activity of mesoporous halloysite nanotubes under heating. Clays Clay Miner 60(6):561–753
96. Hassan-Nejad M, Ganster J, Bohn A, Pinnow M, Volkert B (2009) Bio-based nanocomposites of cellulose acetate and nano-clay with superior mechanical properties. Macromol Symp 280(1):123–129
97. Joo Y, Jeon Y, Lee SU, Sim JH, Ryu J, Lee S, Lee H, Sohn D (2012) Aggregation and stabilization of carboxylic acid functionalized halloysite nanotubes (HNT-COOH). J Phys Chem C116(34):18230–18235
98. Liu M, Guo B, Zou Q, Du M, Jia D (2008) Interactions between halloysite nanotubes and 2, 5-bis (2-benzoxazolyl) thiophene and their effects on reinforcement of polypropylene/halloysite nanocomposites. Nanotechnology 19(20):205709
99. Liu M, Guo B, Du M, Lei Y, Jia D (2008) Natural inorganic nanotubes reinforced epoxy resin nanocomposites. J Polym Res 15(3):205–212
100. Li C, Liu J, Qu X, Guo B, Yang Z (2008) Polymer-modified halloysite composite nanotubes. J Appl Polym Sci 110(6):3638–3646
101. Haroosh HJ, Dong Y, Chaudhary DS, Ingram GD, Yusa S (2013) Electrospun PLA: PCL composites embedded with unmodified and 3-aminopropyltriethoxysilane (ASP) modified halloysite nanotubes (HNT). Appl Phys A 110(2):433–442
102. Riza Erdogan A, Kaygusuz I, Kaynak C (2014) Influences of aminosilanization of halloysite nanotubes on the mechanical properties of polyamide-6 nanocomposites. Polym Compos 35(7):1350–1361
103. Liu C, Luo YF, Jia ZX, Zhong BC, Li SQ, Guo BC, Jia DM (2011) Enhancement of mechanical properties of poly (vinyl chloride) with polymethyl methacrylate-grafted halloysite nanotube. Exp Polym Lett 5(7):591–603
104. Zhang J, Zhang D, Zhang A, Jia Z, Jia D (2013) Poly (methyl methacrylate) grafted halloysite nanotubes and its epoxy acrylate composites by ultraviolet curing method. J Reinf Plast Compos 32(10):713–725
105. Yah WO, Takahara A, Lvov YM (2012) Selective modification of halloysite lumen with octadecylphosphonic acid: new inorganic tubular micelle. JACS 134(3):1853–1859
106. Solomon D (1968) Clay minerals as electron acceptors and/or electron donors in organic reactions. Clays Clay Miner 16:31–39
107. Liu M, Guo B, Du M, Jia D (2008) The role of interactions between halloysite nanotubes and 2, 2′-(1, 2-ethenediyldi-4, 1-phenylene) bisbenzoxazole in halloysite reinforced polypropylene composites. Polym J 40(11):1087–1093
108. Du M, Guo B, Liu M, Jia D (2007) Formation of reinforcing inorganic network in polymer via hydrogen bonding self-assembly process. Polym J 39(3):208–212
109. Du M, Guo B, Liu M, Cai X, Jia D (2010) Reinforcing thermoplastics with hydrogen bonding bridged inorganics. Phys B Condens Matter 405(2):655–662
110. Zhu J, Jia J, Tjong SC (2014) Preparation, structure, and application of carbon nanotubes/bamboo charcoal composite. In: Tjong SC (ed) Nanocrystalline materials: their synthesis-structure-property relationships and applications. Elsevier: London, pp 1–25
111. Zuo S-L, Gao S-Y, Yuan X-G, Xu B-S (2003) Carbonization mechanism of bamboo (phyllostachys) by means of Fourier transform infrared and elemental analysis. J Forestry Res 14(1):75–79

112. Jiang Z-H, Zhang DS, Fei BH, Yue YD, Chen XH (2004) Effects of carbonization temperature on the microstructure and electrical conductivity of bamboo charcoal. New Carbon Mater 19(4):249–253
113. Zhu J, Jia J, Kwong FL, Ng DH, Tjong SC (2012) Synthesis of multiwalled carbon nanotubes from bamboo charcoal and the roles of minerals on their growth. Biomass Bioenerg 36:12–19
114. Cheng H-M, Endo H, Okabe T, Saito K, Zheng GB (1999) Graphitization behavior of wood ceramics and bamboo ceramics as determined by X-ray diffraction. J Porous Mater 6(3):233–237
115. Ye YJ, Zhang ZF (2013) Research progress of the properties and application of bamboo charcoal. Appl Mech Mater 395–396:646–649
116. Asada T, Ishihara S, Yamane T, Toba A, Yamada A, Oikawa K (2002) Science of bamboo charcoal: study on carbonizing temperature of bamboo charcoal and removal capability of harmful gases. J Health Sci 48(6):473–479
117. Gray M, Johnson MG, Dragila MI, Kleber M (2014) Water uptake in biochars: the roles of porosity and hydrophobicity. Biomass Bioenerg 61:196–205
118. Brockhoff SR, Christians NE, Killorn RJ, Horton R, Davis DD (2010) Physical and mineral-nutrition properties of sand-based turfgrass root zones amended with biochar. Agronomy J 102(6):1627–1631
119. You Z, Li D (2014) Highly filled bamboo charcoal powder reinforced ultra-high molecular weight polyethylene. Mater Lett 122:121–124
120. Wu KH, Ting TH, Wang GP, Yang CC, Tsai CW (2008) Synthesis and microwave electromagnetic characteristics of bamboo charcoal/polyaniline composites in 2–40 GHz. Synth Met 158(17–18):688–694
121. Baughman RH, Zakhidov AA, De Heer WA (2002) Carbon nanotubes–the route toward applications. Science 297(5582):787–792
122. Coleman JN, Khan U, Blau WJ, Gun′ko YK (2006) Small but strong: a review of the mechanical properties of carbon nanotube–polymer composites. Carbon 44(9):1624–1652
123. An KH, Kim WS, Park YS, Moon J-M, Bae DJ, Lim SC, Lee YS, Lee YH (2001) Electrochemical properties of high-power supercapacitors using single-walled carbon nanotube electrodes. Adv Funct Mater 11(5):387–392
124. Niu C, Sichel EK, Hoch R, Moy D, Tennent H (1997) High power electrochemical capacitors based on carbon nanotube electrodes. Appl Phys Lett 70(11):1480–1482
125. Zandonella C (2001) Is it all just a pipe dream? Nature 410:734–735
126. Reddy MM, Vivekanandhan S, Misra M, Bhatia SK, Mohanty AK (2013) Biobased plastics and bionanocomposites: current status and future opportunities. Prog Polym Sci 38(10–11):1653–1689
127. Ma P-C, Siddiqui NA, Marom G, Kim JK (2010) Dispersion and functionalization of carbon nanotubes for polymer-based nanocomposites: a review. Compos Part A Appl Sci Manuf 41(10):1345–1367
128. Baur J, Silverman E (2007) Challenges and opportunities in multifunctional nanocomposite structures for aerospace applications. MRS Bull 32(4):328–334
129. Postek MT, Vladár A, Dagata J, Farkas N, Ming B, Wagner R, Raman A, Moon RJ, Sabo R, Wegner TH, Beecher J (2011) Development of the metrology and imaging of cellulose nanocrystals. Meas Sci Technol 22(2):024005
130. Freire CS, Silvestre AJ, Neto CP, Gandini A, Martin L, Mondragon I (2008) Composites based on acylated cellulose fibers and low-density polyethylene: effect of the fiber content, degree of substitution and fatty acid chain length on final properties. Compos Sci Technol 68(15–16):3358–3364
131. Hu K, Gupta MK, Kulkarni DD, Tsukruk VV (2013) Ultra-robust graphene oxide-silk fibroin nanocomposite membranes. Adv Mater 25(16):2301–2307
132. Li C, Adamcik J, Mezzenga R (2012) Biodegradable nanocomposites of amyloid fibrils and graphene with shape-memory and enzyme-sensing properties. Nat Nanotechnol 7(7):421
133. Lawrence BD, Wharram S, Kluge JA, Leisk GG, Omenetto FG, Rosenblatt MI, Kaplan DL (2010) Effect of hydration on silk film material properties. Macromol Biosci 10(4):393–403

134. Chang Y, Yang ST, Liu JH, Dong E, Wang Y, Cao A, Liu Y, Wang H (2011) In vitro toxicity evaluation of graphene oxide on A549 cells. Toxicol Lett 200(3):201–210
135. Ghebaur A, Garea SA, Iovu H (2012) New polymer halloysite hybrid materials—potential controlled drug release system. Int J Pharm 436(1–2):568–573
136. Mallakpour S, Barati A (2012) Application of modified cloisite Na+ with L-phenylalanine for the preparation of new poly (vinyl alcohol)/organoclay bionanocomposite films. Polym Plast Technol Eng 51(3):321–327
137. Mallakpour S, Shahangi V (2013) Bio-modification of cloisite Na+ with chiral L-leucine and preparation of new poly (vinyl alcohol)/organo-nanoclay bionanocomposite films. Synth React Inorg M 43(8):966–971
138. Kaboorani A, Riedl B, Kaboorani B (2013) Ultrasonication technique: a method for dispersing nanoclay in wood adhesives. J Nanomat 2013:341897
139. Mallakpour S, Moslemi S (2012) Dispersion of chiral amino acid organomodified Cloisite Na+ in poly (vinyl alcohol) matrix for designing of novel bionanocomposite films. Prog Org Coat 74(1):8–13
140. Strawhecker K, Manias E (2000) Structure and properties of poly (vinyl alcohol)/Na+ montmorillonite nanocomposites. Chem Mater 12(10):2943–2949
141. Gaaz TS, Sulong AB, Akhtar MN, Kadhum AAH, Mohamad AB, Al-Amiery AA (2015) Properties and applications of polyvinyl alcohol, halloysite nanotubes and their nanocomposites. Molecules 20(12):22833–22847
142. Pavlidou S, Papaspyrides C (2008) A review on polymer–layered silicate nanocomposites. Prog Polym Sci 33(12):1119–1198
143. Ray SS, Okamoto M (2003) Polymer/layered silicate nanocomposites: a review from preparation to processing. Prog Polym Sci 28(11):1539–1641
144. Kojima Y, Usuki A, Kawasumi M, Okada A, Fukushima Y, Kurauchi T, Kamigaito O (1993) Mechanical properties of nylon 6-clay hybrid. J Mater Res 8(5):1185–1189
145. Okamoto M, Morita S, Kotaka T (2001) Dispersed structure and ionic conductivity of smectic clay/polymer nanocomposites. Polymer 42(6):2685–2688
146. Akelah A, Moet A (1996) Polymer-clay nanocomposites: free-radical grafting of polystyrene on to organophilic montmorillonite interlayers. J Mater Sci 31(13):3589–3596
147. Hsu SL-C, Chang K-C (2002) Synthesis and properties of polybenzoxazole–clay nanocomposites. Polymer 43(15):4097–4101
148. Tudor J, Willington L, O'Hare D, Royan B (1996) Intercalation of catalytically active metal complexes in phyllosilicates and their application as propene polymerisation catalysts. Chem Commun 17:2031–2032
149. Sepehr M, Utracki LA, Zheng X, Wilkie CA (2005) Polystyrenes with macro-intercalated organoclay. Part I. Compounding and characterization. Polymer 46(25):11557–11568
150. Lin Y, Ng KM, Chan CM, Sun G, Wu J (2011) High-impact polystyrene/halloysite nanocomposites prepared by emulsion polymerization using sodium dodecyl sulfate as surfactant. J Colloid Interface Sci 358(2):423–429
151. Andrews R, Jacques D, Minot M, Rantell T (2002) Fabrication of carbon multiwall nanotube/polymer composites by shear mixing. Macromol Mater Eng 287(6):395–403
152. Breuer O, Sundararaj U (2004) Big returns from small fibers: a review of polymer/carbon nanotube composites. Polym Compos 25(6):630–645
153. Gorrasi G, Tortora M, Vittoria V, Pollet E, Lepoittevin B, Alexandre M, Dubois P (2003) Vapor barrier properties of polycaprolactone montmorillonite nanocomposites: effect of clay dispersion. Polymer 44(8):2271–2279
154. Pluta M (2004) Morphology and properties of polylactide modified by thermal treatment, filling with layered silicates and plasticization. Polymer 45(24):8239–8251
155. Garcıa-López D, Picazo O, Merino JC, Pastor JM (2003) Polypropylene–clay nanocomposites: effect of compatibilizing agents on clay dispersion. Eur polym J 39(5):945–950
156. Zhou WY, Guo B, Liu M, Liao R, Rabie AB, Jia D (2010) Poly (vinyl alcohol)/halloysite nanotubes bionanocomposite films: Properties and in vitro osteoblasts and fibroblasts response. J Biomed Mater Res, Part A 93(4):1574–1587

157. Lin C-A, An T-C, Hsu Y-H (2007) Study on the far infrared ray emission property and adsorption performance of bamboo charcoal/polyvinyl alcohol fiber. Polym-Plast Technol Eng 46(11):1073–1078
158. Li J, Zhitomirsky I (2009) Cathodic electrophoretic deposition of manganese dioxide films. Colloid Surf A: Physicochem Eng Aspect 348(1–3):248–253
159. Deen I, Pang X, Zhitomirsky I (2012) Electrophoretic deposition of composite chitosan–halloysite nanotube–hydroxyapatite films. Colloid Surf A: Physicochem Eng Aspect 410:38–44
160. Deen I, Zhitomirsky I (2014) Electrophoretic deposition of composite halloysite nanotube–hydroxyapatite–hyaluronic acid films. J Alloy Compound 586:S531–S534
161. Fornes T, Yoon PJ, Hunter DL, Keskkula H, Paul DR (2002) Effect of organoclay structure on nylon 6 nanocomposite morphology and properties. Polymer 43(22):5915–5933
162. Krishnamoorti R, Giannelis EP (1997) Rheology of end-tethered polymer layered silicate nanocomposites. Macromolecules 30(14):4097–4102
163. Raquez J-M, Habibi Y, Murariu M, Dubois P (2013) Polylactide (PLA)-based nanocomposites. Prog Polym Sci 38(10–11):1504–1542
164. Vaia RA, Giannelis EP (1997) Polymer melt intercalation in organically-modified layered silicates: model predictions and experiment. Macromolecules 30(25):8000–8009
165. Krishnamoorti R, Vaia RA, Giannelis EP (1996) Structure and dynamics of polymer-layered silicate nanocomposites. Chem Mater 8(8):1728–1734
166. Wang B, Wang Q, Li L (2014) Morphology and properties of poly (vinyl alcohol)/MMT nanocomposite prepared by solid-state shear milling (S3M). J Macromol Sci Part B 53(1):78–92
167. Mondal D, Mollick MM, Bhowmick B, Maity D, Bain MK, Rana D, Mukhopadhyay A, Dana K, Chattopadhyay D (2013) Effect of poly (vinyl pyrrolidone) on the morphology and physical properties of poly (vinyl alcohol)/sodium montmorillonite nanocomposite films. Prog Nat Sci Mater Int 23(6):579–587
168. Allison PG, Moser RD, Chandler MQ, Caminero-Rodriguez JA, Torres-Cancel K, Rivera OG, Goodwin JR, Gore ER, Weiss CA (2015) Mechanical, thermal, and microstructural analysis of polyvinyl alcohol/montmorillonite nanocomposites. J Nanomater 2015:291248
169. Tian H, Wang K, Liu D, Yan J, Xiang A, Rajulu AV (2017) Enhanced mechanical and thermal properties of poly (vinyl alcohol)/corn starch blends by nanoclay intercalation. Int J Biol Macromol 101:314–320
170. Zhang L, Wang H, Jin C, Zhang R, Li L, Li X, Jiang S (2017) Sodium lactate loaded chitosan-polyvinyl alcohol/montmorillonite composite film towards active food packaging. Innov Food Sci Emerg Technol 42:101–108
171. Hotta S, Paul D (2004) Nanocomposites formed from linear low density polyethylene and organoclays. Polymer 45(22):7639–7654
172. Rooj S, Das A, Heinrich G (2011) Tube-like natural halloysite/fluoroelastomer nanocomposites with simultaneous enhanced mechanical, dynamic mechanical and thermal properties. Eur Polym J 47(9):1746–1755
173. Albdiry M, Yousif B (2013) Morphological structures and tribological performance of unsaturated polyester based untreated/silane-treated halloysite nanotubes. Mater Design 48:68–76
174. Abu Taqa AG, Abu Al-Rub RK, Senouci A, Popelka A, Al-Nuaimi N, Bani-Hani KA (2017) Experimental prediction of the elastic properties of nanocomposite cementitious materials based on nanoindentation measurements. Sci Adv Mater 9(5):830–846
175. Lewis T (2005) Interfaces: nanometric dielectrics. J Phys Part D: Appl Phys 38(2):202
176. Virgilio N, Favis BD, Pépin MF, Desjardins P, L'Espérance G (2005) High contrast imaging of interphases in ternary polymer blends using focused ion beam preparation and atomic force microscopy. Macromolecules 38(6):2368–2375
177. Jesson DA, Watts JF (2012) The interface and interphase in polymer matrix composites: effect on mechanical properties and methods for identification. Polym Rev 52(3):321–354

178. Li Y, Huang Y, Krentz T, Natarajan B, Neely T, Schadler LS (2016) Polymer nanocomposite interfaces: the hidden lever for optimizing performance in spherical nanofilled polymers. In: Mittal KL, Netravali AN (eds) Interface/interphase in polymer nanocomposites. Scrivener Publishing LLC, Beverly, pp 1–69
179. Qu M, Deng F, Kalkhoran SM, Gouldstone A (2011) Nanoscale visualization and multiscale mechanical implications of bound rubber interphases in rubber–carbon black nanocomposites. Soft Matt 7(3):1066–1077
180. Litvinov V, Steeman P (1999) EPDM–Carbon black interactions and the reinforcement mechanisms, as studied by low-resolution 1H NMR. Macromolecules 32(25):8476–8490
181. Pompe G, Mäder E (2000) Experimental detection of a transcrystalline interphase in glass-fibre/polypropylene composites. Compos Sci Technol 60(11):m2159–m2167
182. Brown D, Marcadon V, Mele P, Alberola ND (2008) Effect of filler particle size on the properties of model nanocomposites. Macromolecules 41(4):1499–1511
183. Li Y, Waas AM, Arruda EM (2011) A closed-form, hierarchical, multi-interphase model for composites—derivation, verification and application to nanocomposites. J Mech Phys Solid 59(1):43–63
184. Gu Y, Li M, Wang J, Zhang Z (2010) Characterization of the interphase in carbon fiber/polymer composites using a nanoscale dynamic mechanical imaging technique. Carbon 48(11):3229–3235
185. Appiah K, Wang Z, Lackey W (2000) Characterization of interfaces in C fiber-reinforced laminated C-SiC matrix composites. Carbon 38(6):831–838
186. Dwivedi H, Mathur RB, Dhami TL, Bahl OP, Monthioux M, Sharma SP (2006) Evidence for the benefit of adding a carbon interphase in an all-carbon composite. Carbon 44(4):699–709
187. Van Landingham M, Dagastine RR, Eduljee RF, McCullough RL, Gillespie JW Jr (1999) Characterization of nanoscale property variations in polymer composite systems: 1. Experimental results. Compos Part A Appl Sci Manuf 30(1):75–83
188. Hodzic A, Kim JK, Lowe AE, Stachurski ZH (2004) The effects of water aging on the interphase region and interlaminar fracture toughness in polymer–glass composites. Compos Sci Technol 64(13–14):2185–2195
189. Wright-Charlesworth DD, Peers WJ, Miskioglu I, Loo LL (2005) Nanomechanical properties of self-reinforced composite poly (methyl methacrylate) as a function of processing temperature. J Biomed Mater Res, Part A 74(3):306–314
190. Gao S-L, Mäder E (2002) Characterisation of interphase nanoscale property variations in glass fibre reinforced polypropylene and epoxy resin composites. Compos Part A Appl Sci Manuf 33(4):559–576
191. Gao S-L, Mäder E, Zhandarov SF (2004) Carbon fibers and composites with epoxy resins: topography, fractography and interphases. Carbon 42(3):515–529
192. Yedla S, Kalukanimuttam M, Winter RM, Khanna SK (2008) Effect of shape of the tip in determining interphase properties in fiber reinforced plastic composites using nanoindentation. J Eng Mater Technol 130(4):041010
193. Jafarzadeh S, Claesson PM, Sundell PE, Pan J, Thormann E (2014) Nanoscale electrical and mechanical characteristics of conductive polyaniline network in polymer composite films. ACS Appl Mater Interfaces 6(21):19168–19175
194. Sababi M, Kettle J, Rautkoski H, Claesson PM, Thormann E (2012) Structural and nanomechanical properties of paperboard coatings studied by peak force tapping atomic force microscopy. ACS Appl Mater Interfaces 4(10):5534–5541
195. Pakzad A, Simonsen J, Yassar RS (2012) Gradient of nanomechanical properties in the interphase of cellulose nanocrystal composites. Compos Sci Technol 72(2):314–319
196. Fortunati E, Puglia D, Monti M, Santulli C, Maniruzzaman M, Kenny JM (2013) Cellulose nanocrystals extracted from okra fibers in PVA nanocomposites. J Appl Polym Sci 128(5):3220–3230
197. Widjojo N, Chung TS, Weber M, Maletzko C, Warzelhan V (2011) The role of sulphonated polymer and macrovoid-free structure in the support layer for thin-film composite (TFC) forward osmosis (FO) membranes. J Membr Sci 383(1–2):214–223

198. Knauer KM, Jennings AR, Bristol AN, Iacono ST, Morgan SE (2016) Enhanced surface properties of branched poly (ether sulfone) from semifluorinated polyhedral oligomeric silsequioxanes. ACS Appl Mater Interfaces 8(19):12434–12444
199. Yu Y-H, Lin CY, Yeh JM, Lin WH (2003) Preparation and properties of poly (vinyl alcohol)–clay nanocomposite materials. Polymer 44(12):3553–3560
200. Katti KS, Katti DR, Dash R (2008) Synthesis and characterization of a novel chitosan/montmorillonite/hydroxyapatite nanocomposite for bone tissue engineering. Biomed Mater 3(3):034122
201. Swapna V, Suresh KI, Saranya V, Rahana MP, Stephen R (2015) Thermal properties of poly (vinyl alcohol)(PVA)/halloysite nanotubes reinforced nanocomposites. Int J Plast Technol 19(1):124–136
202. Cavallaro G, De Lisi R, Lazzara G, Milioto S (2013) Polyethylene glycol/clay nanotubes composites. J Therm Anal Calorim 112(1):383–389
203. Schmitt H, Prashantha K, Soulestin J, Lacrampe MF, Krawczak P (2012) Preparation and properties of novel melt-blended halloysite nanotubes/wheat starch nanocomposites. Carbohyd Polym 89(3):920–927

Chapter 2
Materials, Methodology and Characterisation Techniques

Abstract In this chapter, material selection, processing procedures and characterisation techniques used for polyvinyl alcohol (PVA) bionanocomposites reinforced with montmorillonite (MMT) clays, halloysite nanotubes (HNTs) and bamboo charcoals (BCs) were detailed to provide a comprehensive guideline for researchers, materials scientists and engineers, as well as industrial R&D staff to explore the feasibility of such new bionanocomposite material systems for potential use in the near future.

Keywords Bionanocomposites · Polyvinyl alcohol (PVA) · Material selection · Characterisation techniques

2.1 Materials

2.1.1 Polyvinyl Alcohol (PVA)

PVA as a popular water-soluble biopolymer was used in this study, which was supplied by Sigma Aldrich Pty Ltd., Australia, with a material type given by MFCD00081922. Such a selected PVA has a molecular weight of 89,000–98,000 g/mol and the degree of hydrolysis of 99.0–99.8% according to its datasheet for material specification [1].

2.1.2 Nanoparticles

Three different structures and shapes of nanofillers were employed in this work, namely Cloisite 30B montmorillonite (MMT) clays, halloysite nanotubes (HNTs) and bamboo charcoals (BCs). All these three nanofillers are naturally abundant and can be used widely in biomaterial applications. Moreover, the effects of their different structures and shapes on material performance and properties of PVA nanocomposites are of particular interest in this study. 1D platelet-like Cloisite 30B clays comprise –OH groups between their clay interlayer areas, which promotes the insertion of PVA molecules between these layers to form intercalated/exfoliated clay structures.

M. Mousa and Y. Dong, *Multiscaled PVA Bionanocomposite Films*,
https://doi.org/10.1007/978-981-15-8771-9_2

In comparison, HNTs have 2D tubular structures with high aspect ratios and functional groups on their surfaces, which thus makes HNTs closely interact with PVA molecules in PVA/HNT nanocomposites. Finally, BCs possess 3D porous structures with the ability to absorb PVA molecules into such internal pores in order to form both strong mechanical and chemical bondings.

2.1.2.1 Bamboo Charcoals (BCs)

Two different types of bioactive BC fillers were used in this study. The first was denoted as microdiameter bamboo charcoals (MBCs) supplied by Jiangshan Luyi Bamboo Charcoals Co., Ltd., China (molecular weight: 12.01 g/mol and density: 0.2–0.5 g/cm^3) with their particle size being less than 800 nm [2]. The second type was referred to as nanodiameter bamboo charcoals (NBCs) purchased from US Research Nanomaterials, Inc. Co., USA (molecular weight: 12.01 g/mol, density: 0.43 g/cm^3 and particle size less than 100 nm [3]).

2.1.2.2 Clay Nanofillers

Organomodified Cloisite 30B clays with methyl, tallow, bis-2-hydroxyethyl, quaternary ammonium were used in this study, which were supplied by Southern Clay Products, USA. Such clays have an interlayer spacing d_{001} = 1.88 nm, density of 1.98 g/cm^3 and cation exchange capacity (CEC) of 90 mequiv/100 g [4]. Their chemical structure is demonstrated in Fig. 2.1.

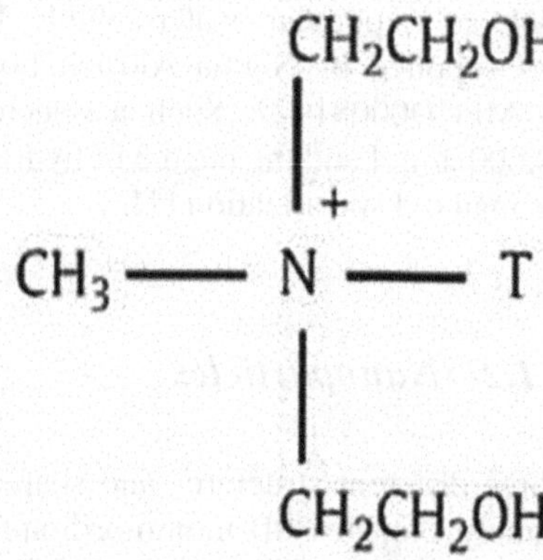

Fig. 2.1 Chemical structure of Cloisite 30B clays

2.1.2.3 Halloysite Nanotubes (HNTs)

HNTs belong to the kaolin group of clay minerals with a chemical formula of $Al_2Si_2O_5(OH)_4{\cdot}nH_2O$ [2]. In this work, HNTs were donated by Imerys Ceramics, Auckland, New Zealand, with their material specification listed in Table 2.1.

2.2 Fabrication of PVA-Based Bionanocomposite Films

PVA/MBC bionanocomposite films were prepared by a solution casting method, as shown in Fig. 2.2. Initially, 5 wt%/v PVA aqueous solution was prepared by dissolving 10 g PVA into 190 ml deionised water via vigorous magnetic stirring at 400 rpm and 90 °C for 3 h until PVA was completely dissolved to prepare a stock solution. MBC aqueous suspension was obtained by using mechanical mixing in deionised water with a rotor speed of 405 rpm at 40 °C for 2 h, which was followed by the ultrasonication (Model ELMA Ti–H–5) at 25 kHz and 40 °C with a power intensity of 70% for 1 h. Subsequently, MBC contents of 0, 3, 5 and 10 wt% were obtained by adding appropriate amounts of PVA. The MBC aqueous suspension was gradually added in a dropwise manner into PVA solutions, simultaneously subjected to mechanical mixing at 405 rpm and 40 °C for 2 h. Further, their mixtures were stirred at 400 rpm and 90 °C for 1 h prior to subsequent sonication for 30 min to achieve uniform dispersion of MBCs. Finally, 20 ml prepared solution was cast on a glass petri dish and allowed to dry in an air-circulating oven at 40 °C for 48 h. Afterwards, PVA/MBC nanocomposite films were stored in a silica gel-containing desiccator before material testing and analysis. PVA/NBC bionanocomposites, PVA/HNT bionanocomposites and PVA/Cloisite 30B clay bionanocomposites were also subjected to the same fabrication procedure in an identical processing condition.

2.3 Characterisation Techniques

The material performance of PVA bionanocomposite films and the effect of different nanofiller contents and shapes on final bionanocomposite films are often evaluated with various characterisation techniques. In this study, morphological structures, mechanical and thermal properties, as well as nanomechanical features were holistically investigated, as shown in Fig. 2.3 and Table 2.2, which are explained in the following subsections.

Table 2.1 Material specification of HNTs [5]

Product	Outer diameter (nm)	Inner diameter (nm)	Length (μm)	SiO_2 (%)	Al_2O_3 (%)	Fe_2O_3 (%)	TiO_2 (%)	Na_2O (%)	MgO (%)
HNTs	40–120	15–100	0.3–1.5	49	34.8	0.35	0.12	0.25	0.15

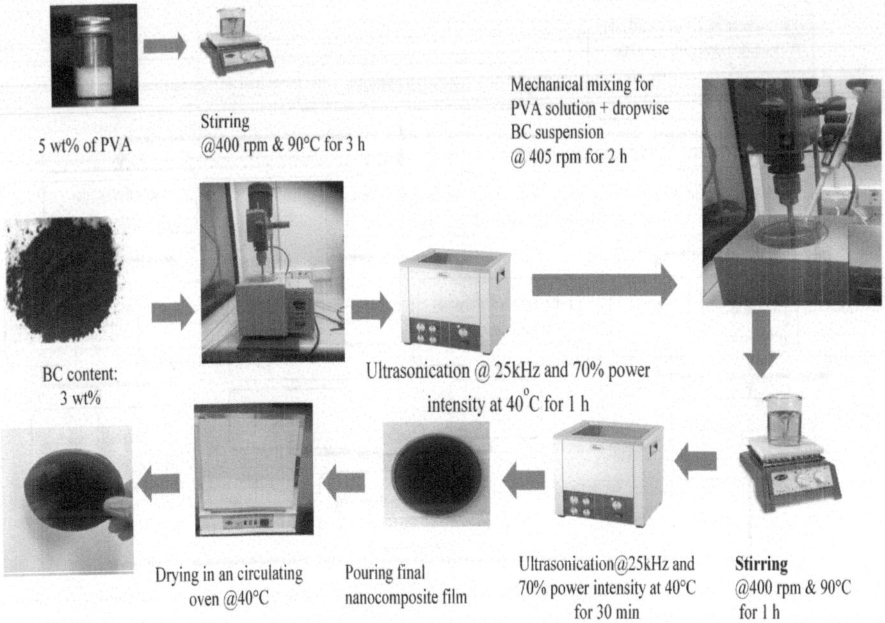

Fig. 2.2 Flowchart for typical fabrication of PVA/MBC nanocomposite films

2.3.1 X-Ray Diffraction (XRD) Analysis

X-ray diffraction analysis is regarded as one of the most popular nondestructive techniques to reveal the structural status of nanocomposites. Moreover, such a technique is widely used to quantify intercalated/exfoliated clay dispersion in MMT-reinforced composites by assessing characteristic peaks of MMTs in measured XRD patterns. For instance, the XRD peak shift to a lower 2θ diffraction angle indicates the expansion of MMT basal spacing to different extent with a clear sign of MMT intercalation. Additionally, the complete disappearance of XRD peaks suggests that full MMT exfoliation may occur, which should be further confirmed with the aid of transmission electron microscopy (TEM). Our XRD analysis was carried out via a Bruker D8 Advance diffractometer, as shown in Fig. 2.4. The X-ray source was Ni-filtered Cu–Kα radiation with the wavelength $\lambda = 0.1541$ nm at the accelerating voltage and current of 40 kV and 40 mA, respectively. The X-ray spectra were recorded at a small-angle 2θ range of 2°–10° and a wide-angle 2θ range of 10°–50° with the scan rate of 0.015°/s. A small-angle range was used to evaluate the intercalation level and dispersion of nanofillers, while a wide-angle range was employed to investigate the crystalline structures of nanofillers and their PVA-based bionanocomposites.

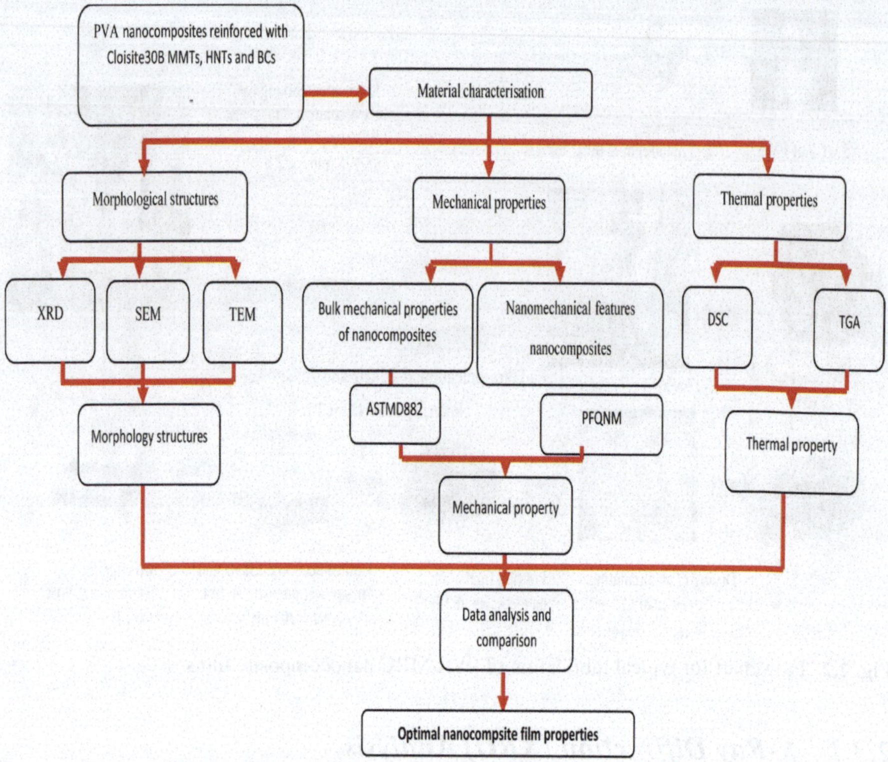

Fig. 2.3 Flowchart for material characterisation of PVA-based bionanocomposite films

2.3.2 *Fourier Transform Infrared (FTIR) Analysis*

FTIR technique is generally used to obtain an infrared spectrum of absorption or emission of a solid, liquid or gas sample based on vibration modes of atoms in a molecule. Such a technique is regarded as the footprint of bonds and atoms for material samples by analysing corresponding absorption peaks located at different frequencies. FTIR was carried out to assess the interaction between PVA matrices and nanofillers. It was recorded in a wavenumber range of 650–4000 cm^{-1} with a resolution of 4 cm^{-1} according to an attenuated total reflectance (ATR) method [6].

2.3.3 *Scanning Electron Microscopy (SEM)*

SEM is one of the most commonly used microscopic techniques to examine the morphology of material fracture surfaces. In SEM, a focused beam with high energy

Table 2.2 Summary of different characterisation techniques

Techniques	Set-up parameters	Material characteristics
XRD	• Ni-filtered Cu–Kα radiation • Accelerating voltage: 40 kV • Accelerating current: 40 mA • 2θ range: 2°–50° • Scan rate: 0.015°/s	• Detecting interlayer distance/spacing of intercalated nanoparticles • Dispersion degree of nanoparticles
FTIR	• Wavenumber range: 650–4000 cm^{-1}	• Component identification and analysis • Interfacial interactions
SEM	• Accelerating voltage: 5 kV • Platinum coating with the layer thickness of 5 nm	• Dispersion degree of nanoparticles • Surface roughness and morphology
DSC	• Temperature range: 35–300 °C • Scan rate: 10 °C/min • Flow rate: 25 ml/min • Argon atmosphere	• Thermal properties in term of glass transition, melting and crystallisation behaviours
TGA	• Temperature range: 35–700 °C • Scan rate: 10 °C/min • Flow rate: 25 ml/min • Argon atmosphere	• Thermal stability in terms of weight loss and thermal decomposition temperatures
Mechanical testing	• Gauge length: 50 mm • Cross-head speed : 10 mm/min	• Young's modulus • Tensile strength • Elongation at break • Tensile toughness
PFQNM in AFM	• RTESPA 525A probes • Spring constant : 200 N/m • Resonant frequency: 525 kHz	• Nanomechanical properties • Nanoroughness • Surface morphological structures

is applied on the sample surfaces to produce various signals involving 2D visual information about sample surface topography and composition [7]. The signals resulting from surface interactions of electron samples are reformatted to generate topographic images. Fracture surface morphology of PVA nanocomposite films was observed with a field emission scanning electron microscope (FE-SEM, Zeiss NEON 40 EsB Cross Beam, as seen in Fig. 2.5) at an accelerating voltage of 5 kV after being coated with platinum (layer thickness: 5 nm) to reduce electric charging effect.

2.3.4 *Differential Scanning Calorimetry (DSC)*

DSC is a thermal analytical technique used to understand the effects of heating and cooling cycles on polymeric behaviour, as well as investigate thermal transitions of polymers and composites [8]. In this study, such a sophisticated technique was employed to study the impact of different nanofiller contents, sizes, shapes and

Fig. 2.4 XRD instrument of a Bruker D8 Advance diffractometer

structures on thermal behaviour of PVA nanocomposite films. DSC was performed on a Mettler Toledo 1 STARe system, as shown in Fig. 2.6. Approximately 8–10 mg film samples were sealed in aluminium pans and heated from 35 to 300 °C at a scan rate of 10 °C/min with a flow rate of 25 ml/min under argon atmosphere. The degree of crystallinity χ_c of PVA matrices in PVA bionanocomposites was calculated as follows:

$$\chi_C(\%) = \frac{\Delta H_m}{w \Delta H_m^O} \times 100\% \tag{2.1}$$

where ΔH_m is the measured melting enthalpy based on DSC data. $\Delta H_m^O =$ 138.6 J/g [9] is the enthalpy of fully crystalline PVA and w is the weight fraction of PVA matrices in corresponding PVA bionanocomposites. Moreover, glass transition temperature (T_g), crystallisation temperature (T_c) and melting temperature (T_m) were determined for PVA bionanocomposite films in DSC data analysis.

2.3.5 *Thermal Gravimetric Analysis (TGA)*

TGA is used to measure the mass loss of samples when the temperature increases over time. In this study, TGA was employed to investigate effects of different nanofillers on thermal degradation of PVA bionanocomposite films. It was also conducted on

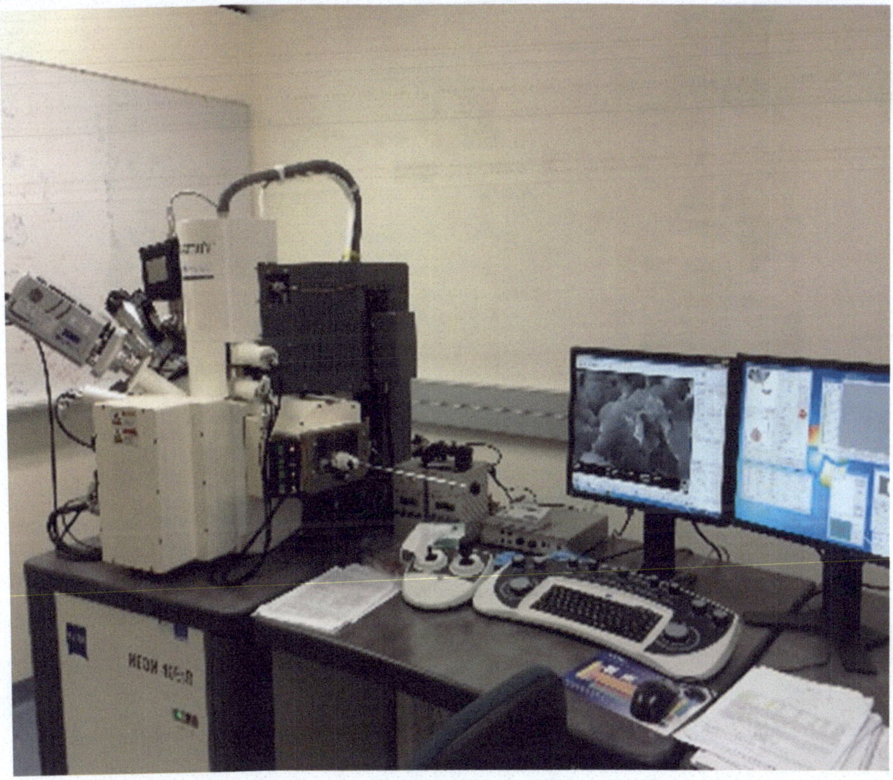

Fig. 2.5 SEM instrument of Zeiss NEON 40 EsB Cross Beam

the same Mettler Toledo 1 STARe system, as shown in Fig. 2.6, in a temperature range from 35 to 700 °C at a scan rate of 10 °C/min with the flow rate of 25 ml/min under the same argon atmosphere.

2.3.6 Mechanical Testing

A universal testing machine Lloyd EZ50 shown in Fig. 2.7 was implemented to carry out conventional tensile tests (gauge length: 50 mm) at the cross-head speed of 10 mm/min at room temperature. For each material batch, six specimens were tested with the mean values and standard deviations being calculated accordingly. The tensile toughness was determined based on tensile energy to break (TEB) according to ASTM D882-02 standard.

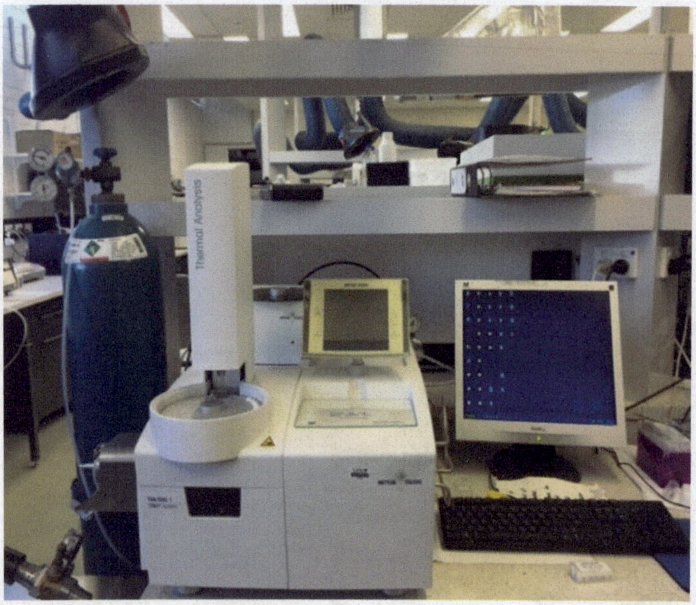

Fig. 2.6 TGA/DSC instrument of Mettler Toledo 1 STARe system

2.3.7 Nanomechanical Measurement

Nanomechanical properties of PVA nanocomposites were quantitatively assessed via atomic force microscopy (AFM) in a peak force quantitative nanomechanical tapping mode (PFQNM) in this study [10]. In this technique, the vertical motion of cantilever is driven at the frequency below its resonance frequency. At each individual tap, nanomechanical properties as well as peak forces are acquired by collecting the force–distance curve at each pixel [11], Fig. 2.8. After that, each force–distance curve is analysed to generate the mapping of material properties at the same resolution, which is known as the height image [11]. During the scanning process, the feedback loop of tapping mode can control the maximum force (i.e. peak force) on the tip so that material sample and AFM tip can be protected from any damage. During the data acquisition process, cantilever deflection and position are converted into a typical force–distance curve. Furthermore, the deflection sensitivity as the representation of deflected distance of the cantilever [10] should be measured accordingly as the requirement for the acquisition of force–distance data. In particular, the determination of deflection sensitivity is based upon the voltage change in the photo detector (nm/V), which can be further converted into the distance (nm) along with the force obtained by multiplying the cantilever deflection with the spring constant [10].

A wide range of material properties such as adhesion, modulus, deformation and energy dissipation can be reported by analysing the force–distance data. The elastic modulus was determined by fitting Derjaguin–Muller–Toropov (DMT) model [13]

Fig. 2.7 Lloyd EZ50 universal testing machine for conventional tensile tests

to initial retrace curve portion, as illustrated in Fig. 2.8 (see the green dash line). This model was developed at the low adhesion force between AFM tip with a small tip-end radius and material sample as opposed to the compliance counterpart.

The load force on the cantilever F_{Lc} can be given in the following equation [13]

$$F_{Lc} = \frac{4}{3}E^{*}\sqrt{R(z - z_o)^3} + F_{adh} \tag{2.2}$$

where E^{*} is the reduced Young's modulus, R is the tip-end radius, $z - z_o$ is the difference between current piezo position z and original position z_0, while F_{adh} is the adhesion force. The sample modulus E_s can be calculated based on E^{*} and elastic modulus of AFM tip E_{tip}, as well as Poisson's ratios of the sample and AFM tip (i.e. v_s and v_{tip}), respectively, according to the following equation [13]:

$$E^{*} = \left[\frac{1 - v_s^2}{E_s} + \frac{1 - v_{tip}^2}{E_{tip}}\right]^{-1} \tag{2.3}$$

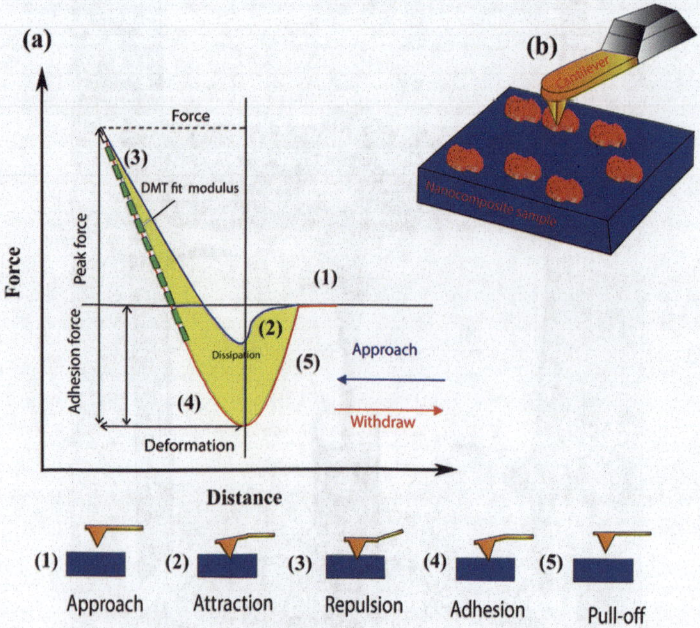

Fig. 2.8 **a** Schematic diagram of force versus tip–sample separation obtained from AFM tapping on the sample surface where the separation is calculated from z piezo position and cantilever deflection. Blue and red curves denote loading and unloading portions, respectively, and green dash line is obtained by DMT fitting. The minimum force in the withdraw curve is used for mapping adhesion force. Points from (1) to (5) represent tip–sample interactions throughout 0.5 ms. **b** Schematic diagram of nanocomposite sample tapping process [12]

Adhesion force represents the attraction force between probe tip and sample surface, which can be directly identified as the minimum force illustrated in Fig. 2.8. The energy dissipation in a cycle of interaction can be given below [10]

$$W = \int \overline{F} \mathrm{d}\bar{z} = \int_{0}^{T} \bar{F} \bar{v} \mathrm{d}t \tag{2.4}$$

where $\overline{F}$ is an interaction force vector and $\mathrm{d}\bar{z}$ is a displacement vector. Because the velocity vector $\bar{v}$ changes its direction at each half cycle, the integration becomes zero when loading and unloading curves appear to be coincident. Moreover, there is no hysteresis phenomenon above zero loads between repulsive portions of loading–unloading curve, which is associated with very low energy dissipation. In such a case, the work of adhesion represents a dominant contributor to energy dissipation. The associated deformation represents the difference in separation from zero cantilever force to peak forces [10].

In this study, a Bruker Dimension Fastscan AFM system was employed to measure peak force quantitative nanomechanical properties and acquire single force–distance curve under the ambient condition. Moreover, RTESPA 525A probes with a nominal spring constant of 200 N/m, a nominal tip radius of 8 nm and a nominal resonant frequency of 525 kHz were utilised for the direct measurement of corresponding nanomechanical properties of PVA bionanocomposite films. Before each measurement, it was confirmed that the deflection sensitivity was calibrated by obtaining a force curve based on the stiff sapphire-12 surface. Later on, a thermal tuning method [14] was applied to determine the spring constant, which was considered as one of the most accurate methods for detecting the spring constant by measuring cantilever mechanical responses to thermal agitation via the Brownian motion of encompassing fluid molecules. AFM imaging analysis was undertaken with the TESPA probe at the nominal spring constant of 40 N/m with a tip radius of 8 nm. The image scan rate was kept at 2 Hz with a digital pixel resolution by 256 × 256. AFM topographic images were the first-order flattened via *Flatten* command in Burker Nanoscope 1.5 software, which was used to remove unwanted features resulting from the vertical (Z) scanner such as noise, bow and tilt.

2.4 Summary

- PVA with the hydrolysis degree of 99.99% was selected as a water-soluble biopolymer with three different types of nanofillers. BCs were chosen because of their 3D irregular shape and porous structures for the absorption of PVA molecules. Additionally, Cloisite 30B clays were employed due to their 1D platelet-like shape with a relatively large interlayer spacing as well as functional groups to facilitate the close interaction with PVA molecular chains. HNTs have been chosen in that such nanofillers possess 2D tubular structures with alumina and silica groups on HNT surfaces, which assists in the good bonding between HNTs and PVA matrices in PVA/HNT bionanocomposites.
- Solvent casting method was used in manufacturing PVA bionanocomposites. The combination process of magnetic stirring, mechanical mixing and ultrasonication is essential for the fabrication of PVA bionanocomposites. The only exception is that no chemical solvent apart from water was used, and such bionanocomposite films were kept in a silica gel-containing desiccator to avoid the moisture effect.
- SEM, XRD and FTIR analyses were carried out to investigate morphological structures of PVA bionanocomposites films. Conventional tensile tests were performed to determine their bulk mechanical properties, while nanomechanical features were evaluated by PFQNM.
- Thermal stability and thermal properties of all bionanocomposites were determined via TGA and DSC, respectively.

References

1. Mousa M, Dong Y, Davies IJ (2018) Eco-friendly polyvinyl alcohol (PVA)/bamboo charcoal (BC) nanocomposites with superior mechanical and thermal properties. Adv Compos Mater 27(5):499–509
2. Material data sheet of bamboo charcoals powder from Jiangshan Luyi Bamboo Charcoals Co., Ltd., China (https://cnjslyzt.en.alibaba.com/)
3. Material data sheet of bamboo charcoals powder from US Research Nanomaterials, Inc. Co., USA (https://us-nano.lookchem.com/)
4. Mousa MH, Dong Y, Davies IJ (2016) Recent advances in bionanocomposites: preparation, properties, and applications. Int J Polym Mater Polym Biomater 65(5):225–254
5. Haroosh HJM (2014) Investigating novel biopolymeric nanostructures for drug delivery. Ph.D. thesis, Curtin University, Perth, Australia
6. Chittur KK (1998) FTIR/ATR for protein adsorption to biomaterial surfaces. Biomaterials 19(4–5):357–369
7. Bozzola JJ (2001) Electron microscopy. Encyclopedia of life sciences. Wiley, Chichester, pp 48–72
8. Menczel JD, Judovits L, Prime RB, Bair HE, Reading M, Swier S (2009) Differential scanning calorimetry (DSC). In: Menczel JD, Prime RB (eds) Thermal analysis of polymers: fundamentals and applications. Wiley, Hoboken, pp 7–239
9. Ma HL, Zhang L, Zhang Y, Wang S, Sun C, Yu H, Zeng X, Zhai M (2016) Radiation preparation of graphene/carbon nanotubes hybrid fillers for mechanical reinforcement of poly (vinyl alcohol) films. Radiat Phys Chem 118:21–26
10. Pittenger B, Erina N, Su C (2010) Quantitative mechanical property mapping at the nanoscale with PeakForce QNM. Application Note Veeco Instruments Inc., pp 1–12
11. Van Der Werf KO, Putman CA, de Grooth BG, Greve J (1994) Adhesion force imaging in air and liquid by adhesion mode atomic force microscopy. Appl Phys Lett 65(9):1195–1197
12. Mousa M, Dong Y (2018) Elastic behavior of nanophases in polyvinyl alcohol (PVA)/bamboo charcoal (BC) nanocomposite films. Front Mater 5:44
13. Derjaguin BV, Muller VM, Toporov YP (1975) Effect of contact deformations on the adhesion of particles. J Colloid Interface Sci 53(2):314–326
14. Ohler B (2007) Practical advice on the determination of cantilever spring constants. Bruker Application Note# 94

Chapter 3
PVA/BC Bionancomposite Films with Particle Size Effect

Abstract The use of conventional carbon-based nanofillers such as carbon nanotubes (CNTs) and graphene sheets can be hindered to a certain extent when considering a high level of nanotoxicity via human inhalation and less cost-effectiveness for CNTs while the agglomeration of graphene sheets in polymer matrices. As such, bamboo charcoals (BCs) are suggested as alternative ecofriendly and sustainable carbon-based particles with good affinity to polyvinyl alcohol (PVA) as one of most popular water-soluble biopolymers in order to prepare PVA/BC bionanocomposites with superior material properties. In this chapter, two different types of BC particles, namely microdiameter bamboo charcoals (MBCs) and nanodiameter bamboo charcoals (NBCs), were successfully fabricated using a solution casting method. Nanofiller reinforcement effect was investigated with respect to BC particle size and dispersion, morphological structures and interfacial interactions between BCs and PVA matrices in such bionanocomposites. Overall, the addition of NBCs gave rise to much higher mechanical properties of PVA/NBC bionanocomposites as opposed to PVA/MBC counterparts. The maximum increases in tensile moduli of bionanocomposites were achieved up to 123 and 100% with the inclusion of 10 wt% NBCs and MBCs, respectively. Furthermore, corresponding tensile strengths were enhanced by 110 and 72% when incorporated with 3 wt% NBCs and MBCs accordingly, as compared with that of PVA. It is convincing that associated particle dispersion states within PVA matrices, as well as interfacial interaction between BCs and PVA matrices play a leading role in the remarkable property enhancement in PVA/BC bionanocomposites, which can be potentially used in material packaging and biomedical applications.

Keywords Polyvinyl alcohol (PVA) · Bamboo charcoals (BCs) · Bionanocomposites · Nanostructures · Mechanical properties · Atomic force microscopy (AFM)

M. Mousa and Y. Dong, *Multiscaled PVA Bionanocomposite Films*,
https://doi.org/10.1007/978-981-15-8771-9_3

3.1 Introduction

The morphology of bamboo charcoals (BCs) generally demonstrates a wide range of internal pore distribution from less than 1 nm to 1 μm [1]. The walls of basic units within BCs, known as parenchyma, are very rough despite their smooth outer surfaces, which makes entire particles quite hard. With a large number of roughly walled pores inside BCs, polymeric chains tend to easily penetrate into internal BC pores [2] to form strong mechanical bonding in addition to the existing hydrogen bonding of pores. Hence, mechanical properties of resulting bionanocomposites can be improved accordingly owing to effective interfacial bonding between nanofillers and polymer matrices. Nevertheless, the selection of polymer matrices also plays a leading role in the manufacture of bionanocomposites in addition to the inclusion of BCs. Polymeric behaviour in a porous medium is associated with capillary forces so that polymeric chains can be further prevented from entering into internal pores [2]. According to the Laplace theory [2, 3], the strength of capillary forces depends on surface chemistry and physical properties of polymers. When hydrophilic media are selected, applied positive capillary pressures drive polymeric chains into BC pores to form both effective mechanical and chemical bondings, which is the reason why PVA as a base polymeric material has been chosen in this study. On the contrary, the surfaces of hydrophobic media generate negative capillary pressures to hinder the penetration of polymeric chains into BC pores, leading to phase separation in a bionanocomposite system. Such a mechanism can be interpreted as typical debonding effect widely observed in polymer/BC composites such as polylactic acid (PLA)/BC composites [4], ultrahigh molecular-weight polyethylene (UHMWPE)/BC composites [5] and polyaniline (PANI)/BC composites [6].

3.2 BC Particle Characterisation

3.2.1 BC Composition and Surface Area

BCs generally comprise the elements of carbon, oxygen, hydrogen, nitrogen and small quantities of ash [7]. Since the carbon content is a function of many material processing parameters such as pyrolysis temperature, moisture content and composition of biomass [8], it is regarded as a key factor to determine the charcoal quality. As seen from Table 3.1, the carbon content of NBCs appears to be relatively high when compared with that of MBCs (i.e. 84.18% vs. 80.04%), which is in good accordance with previous results obtained by Li et al. [7]. As such, BCs can be considered as good carbon-based fillers for effective reinforcements in bionanocomposite systems. Oxygen, hydrogen and nitrogen contents for both NBCs and MBCs (O%: 5–10%, H%: 2.2–2.5% and N%: 0.5–0.7%) become less pronounced accordingly.

Table 3.1 Chemical composition and surface areas of NBCs and MBCs [9]

BC type	C (%)	H (%)	N (%)	O (%)	Ash (%)	S_{BET} ($m^2\ g^{-1}$)	S_{mic} ($m^2\ g^{-1}$)	S_{ext} ($m^2\ g^{-1}$)	V_{mic} ($cm^3\ g^{-1}$)	D_{pore} (nm)
NBC	84.18	2.22	0.66	4.6	8.34	624.81	415.46	209.35	0.17	2.87
MBC	80.04	2.54	0.52	9.86	6.93	527.41	390.84	136.57	0.16	4.14

O (%) = 100-(C + H + N + Ash), S_{BET} = surface area, S_{mic} = smicropore area, S_{ext} = external surface area, V_{mic} = micropore volume and D_{pore} = pore diameter

Moreover, surface area and pore volume tend to be the other two important controlling factors in effective BC reinforcements. As mentioned previously by Das et al. [10], pore volume and pore number in BCs could enhance the absorption capacity of polymer matrices to allow for their sufficient interfacial bonding with BCs. All three measured major surface areas including Brunauer–Emmett–Teller (BET) surface area (S_{BET}), micropore area (S_{micro}) and external surface area (S_{ext}) appear to be consistently higher for NBCs when compared with those of MBCs despite comparable micropore volume (V_{micro}) and relatively low pore diameter (D_{pore}) of NBCs. Such results imply that NBCs have much better adsorption capability towards surrounding media, and thus can yield more efficient load transfer from polymer matrices to fillers, which is in good agreement with previous work from She et al. [11]. The increase in NBC surface areas may be attributed to high carbonisation temperature ($\geq$1300 °C) and small particle size [11, 12].

3.2.2 Particle Size and Elastic Modulus

Topographical analysis in relation to morphological structures of as-received NBCs and MBCs was conducted by means of peak force quantitative nanomechanical mapping (PFQNM) via atomic force microscopy (AFM), Fig. 3.1. After the holistic evaluation of 1356 near-spherical NBCs and 871 MBCs, it has been statistically determined that average particle diameters are 69.43 and 406.8 nm for NBCs and MBCs, respectively. In addition, the average thickness of NBCs is about 6 nm as compared to 54.4 nm for that of MBCs. Figure 3.1a, c as well as Fig. 3.1b, d display typical cases of NBCs and MBCs deposited on steel substrates from aqueous dispersion at different magnifications, respectively. It is clearly shown that NBCs are more uniformly dispersed with disc-like particle shapes and smaller particle sizes when compared with sharp-edge MBCs in a 2D view. The variations of particle size and thickness visually confirm different surface areas listed in Table 3.1.

Young's moduli of NBCs and MBCs are essentially used for the prediction of elastic moduli of corresponding PVA/BC bionanocomposites. However, so far, there has been no published literature to report such important parameters in an experimental manner. Nanomechanical properties of NBCs and MBCs have also been determined by AFM using PFQNM to quantitatively measure the stiffness of BCs as nanoreinforcements. Five different zones of interest on NBCs and MBCs were selected in typical mapping images, illustrated in Figs. 3.2 and 3.3, respectively.

A set of elastic modulus values was estimated according to Derjaguin–Mueller–Toporov (DMT) model [13, 14] for each targeted zone based on corresponding modulus mapping distributions. With the aid of mapping Gaussian distribution curves, elastic moduli of NBCs and MBCs were determined to be 84.5 $\pm$ 3.6 GPa and 80.65 $\pm$ 2.1GPa, respectively. Such results suggested that particle size effects are minor on elastic moduli of NBCs and MBCs. When compared with those of other carbon-based fillers, elastic moduli of NBCs and MBCs are higher than that of T300 carbon fibres [15], which is in range of 20–40 GPa as well as higher than the

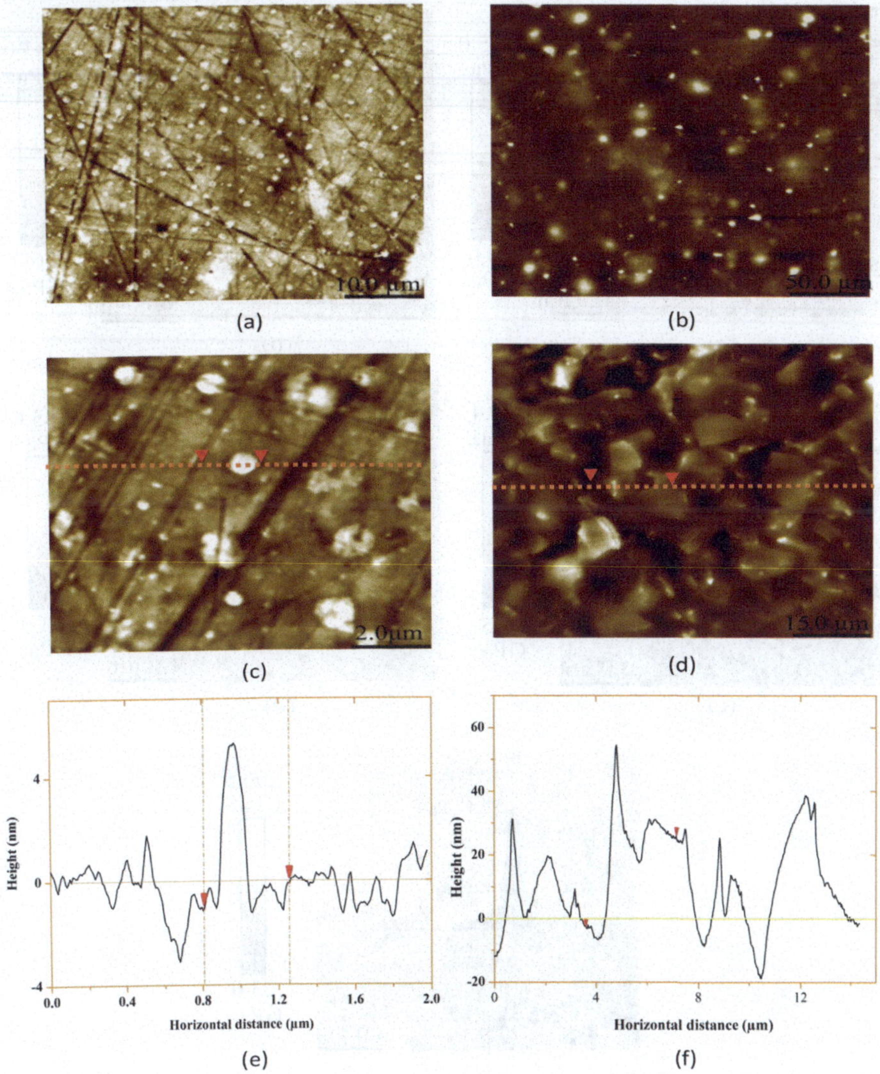

Fig. 3.1 AFM characterisation of individual BC particles: 2D mapping images of NBCs **a**, **c** as well as MBCs **b**, **d** at different magnifications deposited on steel substrates in aqueous solutions. **e**, **f** are height profiles of NBCs and MBCs, respectively [9]

modulus of carbon nanoparticles at 43.89 GPa [16]. However, the elastic modulus of BCs determined in this study is still much lower than those of graphene oxides (GOs) between 200 and 250 GPa [17] and multi-walled carbon nanotubes (MWNTs) at 0.9 TPa [18]. However, GO sheets can have very limited use with a higher tendency to form agglomerates [19], as well as their detrimental effect on human cells, as evidenced by decreasing A549 cell viability at high GO content levels [20]. On the

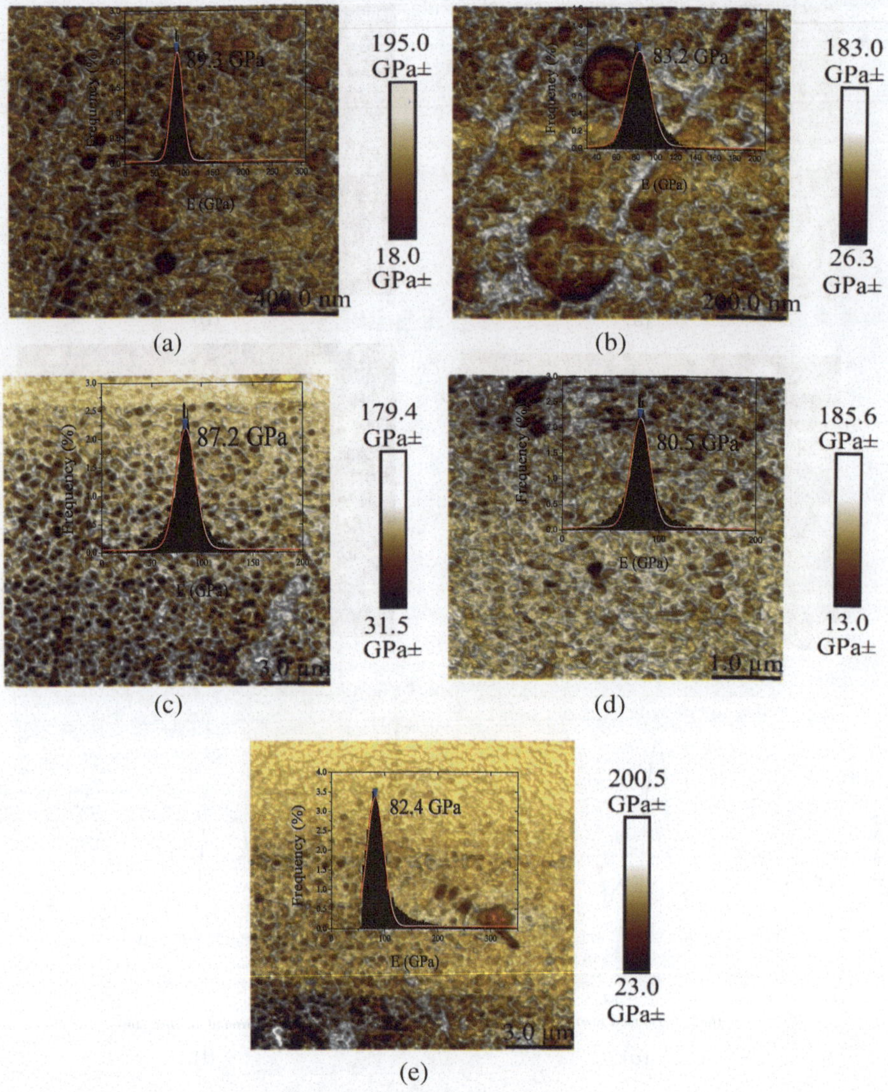

Fig. 3.2 Histograms of elastic modulus distribution of AFM mapping images: **a–e** are five different typical zones of interest for NBCs. Elastic modulus mapping distribution is fitted with Gaussian distribution curves for data histograms [9]

other hand, CNTs possess significant drawbacks with high material cost and nanotoxicity since their accumulation in cytoplasm has the ability of destroying human cells under certain inhalational conditions [21]. As such, BCs are regarded as favourable carbon-based nanofillers to potentially replace popular GOs and CNTs in the future.

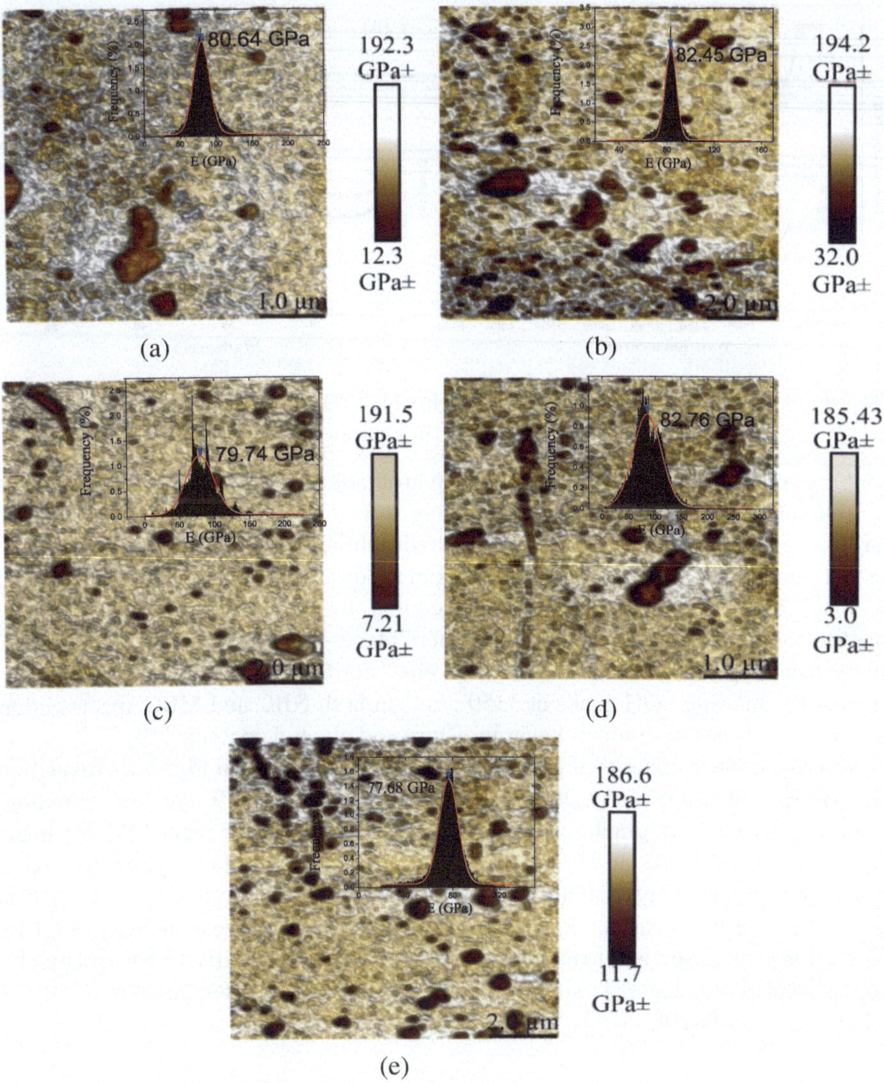

Fig. 3.3 Histograms of elastic modulus distribution of AFM mapping images: **a**–**e** are five different typical zones of interest for MBCs. Elastic modulus mapping distribution is fitted with Gaussian distribution curves for data histograms [9]

3.2.3 FTIR and XRD Analyses

FTIR spectra are shown in Fig. 3.4a to investigate the surface functionality of as-received MBCs and NBCs. The peaks at approximately 2439.6 and 2417.5 cm^{-1} are assigned to the $C\equiv H$ stretching [22]. Whereas, the peaks about 1567.4 and

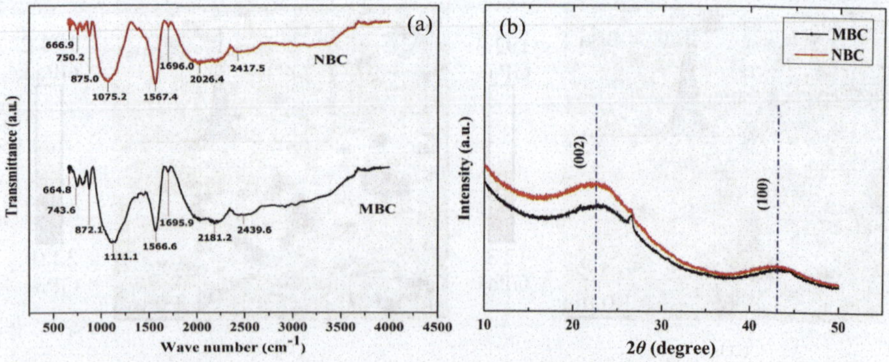

Fig. 3.4 As-received NBCs and MBCs: **a** FTIR spectra [9] and **b** XRD patterns

1566.6 cm^{-1} represent C=C vibration in an aromatic system [11]. The peak spectra at 1696 and 1695.9 cm^{-1} are corresponding to C=O band primarily for ionisable carboxyl groups as an indicator of surface hydrophilicity [23]. The bands existing at 1111.1 and 1075.2 cm^{-1} are ascribed to axial deformation of C–O band. Moreover, the bands at 872.1 and 875 cm^{-1} are referred to as C–H bending (in plane) and C–H bending (out of plane), respectively. The out of plane –OH bending is also designated by the band appearance at 750.2 and 743.6 cm^{-1} for NBCs and MBCs, respectively. Finally, the missing –OH peaks at 3350 cm^{-1} in both NBC and MBC spectra infer that both BC types have much lower moisture and alcohol contents [10].

Moreover, XRD patterns of MBCs and NBCs are depicted in Fig. 3.4b, revealing the existence of two broad peaks. The board peaks at $2\theta = 22.9°$ are corresponding to those sharp peaks of graphite assigned to the (002) diffraction plane [24]. Besides, the second broad peaks at $2\theta = 43.6°$ characterised 2D in-plane symmetry (101) along with graphene layers. Those board bands confirm that both NBCs and MBCs belong to amorphous carbon. Moreover, NBCs exhibit the highest intensity of (101) diffraction peak as opposed to that of MBCs, which means that NBCs have a relatively high graphitisation degree arising from a higher carbonisation temperature of NBCs relative to that of MBCs [25].

3.3 Characterisation and Properties of PVA/BC Bionanocomposites

3.3.1 FTIR and XRD Spectra

FTIR spectra of embedded NBCs and MBCs within PVA matrices in PVA/BC bionanocomposites are illustrated in Fig. 3.5b, c. The change of wave number for –OH and –C–OH stretching bands is well known to be sensitive to hydrogen bonding

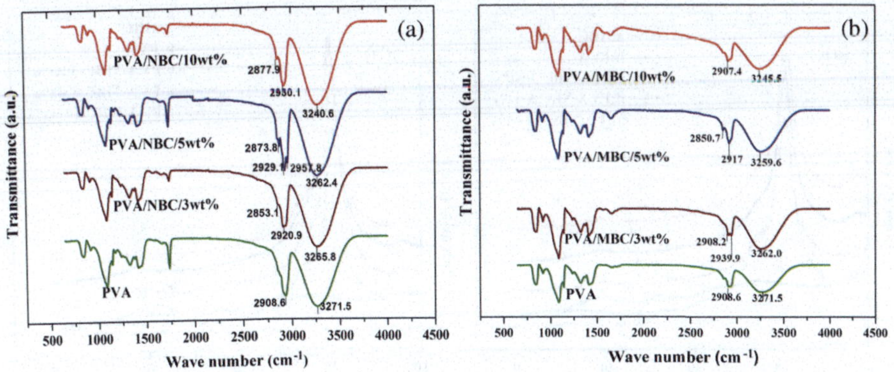

Fig. 3.5 FTIR spectra of **a** PVA/NBC bionanocomposites and **b** PVA/MBC bionanocomposites [9]

[9]. Apparently, the originally wide band at approximately 3271.5 cm^{-1} for neat PVA is assigned to a strong hydroxyl band for free and hydrogen bonded alcohols. With increasing NBC and MBC contents from 0 to 10 wt% in PVA/BC bionanocomposites, this band peak shifts to lower wave numbers at 3240.6 and 3245.5 cm^{-1}, respectively. This finding can be related to large quantities of hydroxyl groups in PVA molecules [9], as well as strong adhesion of NBC or MBC surfaces to PVA matrices. This phenomenon facilitates the occurrence of hydrogen bonds intertwined at PVA/BC interfaces with a broad O–H band. Such a variation associated with –OH stretching vibration confirms the formation of hydrogen bonds, which is similar to those between PVA matrices and graphene sheets in PVA/graphene nanocomposites [26] and between PVA matrices and BCs in PVA/BC nanocomposites [27]. When compared with the PVA spectrum, new bands at 2853 and 2874 cm^{-1} have been detected at the NBC contents of 3 and 5 wt%, respectively, which are assigned to –CH_2– asymmetric and symmetric stretchings. The appearance of such new bands and their band-increasing tendency demonstrate that NBCs have stronger chemical bonding effect with PVA molecular chains [28]. At the BC content of 5 wt%, a new band at 2957.8 cm^{-1} assigned to –CH_2– stretching takes place for PVA/NBC bionanocomposites along with other significantly increased peaks. However, overall relatively low FTIR peaks are more manifested for PVA/MBC bionanocomposites in a wavenumber range of 2800–3300 cm^{-1}. This phenomenon may lie in a higher level of NBC interactions with PVA molecular chains as opposed to MBCs to form much stronger interfacial areas owing to relatively high surface areas of NBCs, previously reported in Table 3.1.

XRD patterns for PVA/NBC bionanocomposites and PVA/MBC bionanocomposites are shown in Fig. 3.6. Pure PVA films reveal the highest diffraction peak at $2\theta = 19.7°$ corresponding to the (101) total crystalline phase of PVA and the second peak at $2\theta = 40.4°$ in relation to semicrystalline structures of PVA [29]. Moreover, XRD patterns for PVA/NBC bionanocomposites in Fig. 3.6a and PVA/MBC

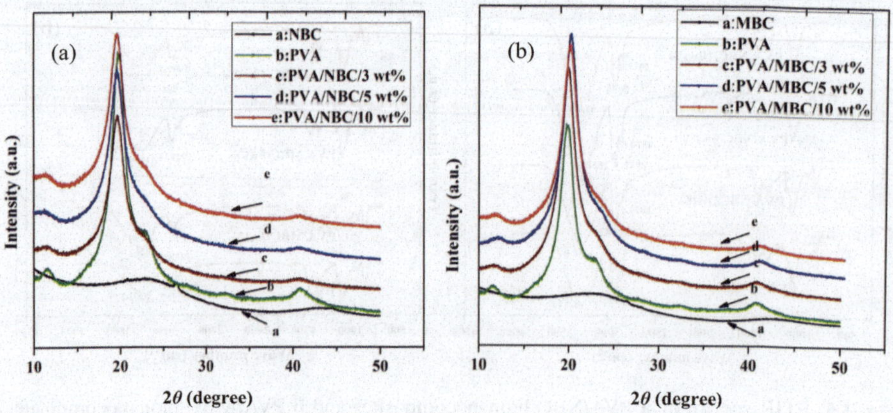

Fig. 3.6 XRD patterns of **a** PVA/NBC bionanocomposites [30] and **b** PVA/MBC bionanocomposites [27]

bionanocomposites, as seen in Fig. 3.6b, demonstrate a peak shift to smaller diffraction angles. Moreover, XRD patterns of bionanocomposites only display PVA peaks, while the diffraction peaks of MBCs and NBCs disappear. Such a finding indicates that PVA molecules can cover nanoparticles and thus make active bonding interactions inside these pores because BC pores possess 'mechanical anchoring' mechanism [9] when interacting with PVA molecular chains. This phenomenon means that a mechanical interlocking phenomenon takes place due to the existence of PVA molecular chains inside BC surfaces [9]. Similar results have also been observed in PVA/5 wt% GO nanocomposites [31].

3.3.2 Mechanical Properties

As illustrated in Fig. 3.7a and Table 3.2, tensile moduli of PVA bionanocomposites reinforced with NBCs and MBCs increase significantly in a monotonic manner with increasing the BC content. The maximum moduli of bionanocomposites taking place at 4.63 and 4.16 GPa have been achieved with 10 wt% NBC and MBC inclusions, respectively, which are equivalent to modulus increases by 123 and 100%, as opposed to that of neat PVA at 2.08 GPa.

On the other hand, tensile strengths of PVA/NBC bionanocomposites and PVA/MBC bionanocomposites demonstrate initial enhancements up to 110 and 72% when the BC content increases from 0 to 3 wt%. Beyond 3 wt% BC inclusions, tensile strengths of both bionanocomposites tend to decline until they reach the lowest strength levels of 96.34 and 80.69 MPa, respectively, at the BC content of 10 wt%.

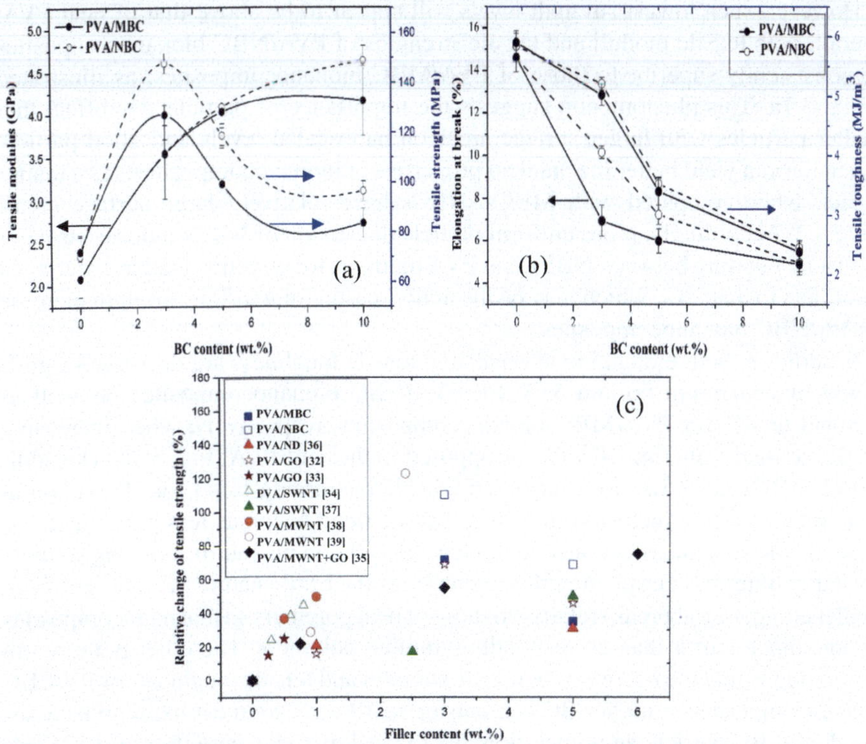

Fig. 3.7 Mechanical properties of PVA/NBC bionanocomposites and PVA/MBC bionanocomposites at different NBC and MBC contents: **a** tensile modulus and tensile strength, **b** elongation at break and tensile toughness and **c** relative change of tensile strength versus filler content [9]

Table 3.2 Mechanical properties of PVA/MBC bionanocomposites and PVA/NBC bionanocomposites [9]

Material type	Tensile modulus (GPa)	Tensile strength (MPa)	Elongation at break (%)	Toughness (MJ/m^3)
PVA	2.08 ± 0.21	70.32 ± 2.9	14.60 ± 0.65	5.84 ± 0.21
PVA/MBC/3 wt%	3.54 ± 0.52	120.64 ± 4.3	6.93 ± 0.72	4.86 ± 0.20
PVA/MBC/5 wt%	4.03 ± 0.39	95.06 ± 3.1	5.98 ± 0.58	3.1 ± 0.25
PVA/MBC/10 wt%	4.16 ± 0.41	80.69 ± 4.4	4.85 ± 0.47	1.96 ± 0.19
PVA/NBC/3 wt%	3.55 ± 0.16	147.94 ± 3.9	10.14 ± 0.29	5.2 ± 0.11
PVA/NBC/5 wt%	4.09 ± 0.25	118.73 ± 4.8	7.22 ± 0.45	3.52 ± 0.11
PVA/NBC/10 wt%	4.63 ± 0.18	96.34 ± 4.1	4.96 ± 0.42	2.45 ± 0.13

However, such lowest strength levels still appear to be above that of neat PVA. Overall, both tensile moduli and tensile strengths of PVA/NBC bionanocomposites are consistently superior to those of PVA/MBC bionanocomposites, as illustrated in Fig. 3.7a. This phenomenon suggests the importance of particle size effect that smaller particles with larger surface areas on nanoscaled levels and good particle dispersion can yield better mechanical properties of bionanocomposites. As a matter of fact, when compared with MBCs, NBCs have relatively large surface areas, Table 3.2. Accordingly, more uniform particle dispersion of NBCs induces stronger interfacial bonding between NBCs and PVA matrices for effective load transfer from nanofillers to matrices, which may be the major reason for a further strength increase in PVA/NBC bionanocomposites.

In addition, both elongation at break and tensile toughness are decreased significantly by maximum 66 and 58% for PVA/NBC bionanocomposites, as well as 66.8 and 66.4% for PVA/MBC bionanocomposites, respectively, when increasing the BC content from 0 to 10 wt%, as opposed to those of PVA, Fig. 3.7b. The addition of MBCs gives rise to a more remarkable decreasing trend for elongation at break relative to the inclusion of NBCs. Such a trend becomes less pronounced as far as tensile toughness is concerned. Generally, the inclusion of more rigid fillers inevitably improves elastic moduli and enhances the brittle nature (as reflected from smaller elongation at break) relative to those of neat polymers so that nanocomposites become much stiffer than corresponding unfilled polymers. The other point worth mentioning is that the decreases in tensile strength and tensile toughness of PVA/BC bionanocomposites at higher BC contents up to 10 wt% are most likely to be associated with BC particle agglomeration due to weak particle–particle van der Waals interactions, thus resulting in high stress concentration zones prone to mechanical failure.

Figure 3.7c demonstrates the comparison of tensile strengths of PVA/NBC bionanocomposites and PVA/MBC bionanocomposites with other PVA nanocomposites available in previous literatures [32–39]. Xu et al. [32] reported that the tensile strength of PVA was increased by 70% with the inclusion of 3 wt% GOs due to strong interfacial bonding between PVA and GOs according to their XRD analysis. Arao et al. [33] stated that the incorporation of 0.25 wt% of few-layer-graphene (FLG) into PVA matrices enhanced tensile strength of PVA/FLG nanocomposites by 15% because uniform FLG dispersion enabled to form stronger interfacial bonding with PVA matrices. Furthermore, Liu et al. [34] functionalised SWNT surfaces with multiple hydrogen bonding groups to improve SWNT dispersion in PVA/SWNT nanocomposites. Their tensile strength was found to be improved by 44.6% at the SWNT content of 0.8 wt%, when compared to that of neat PVA. The use of hybrid nanofillers was also presented by Li et al. [35] to achieve increases in tensile strengths of PVA nanocomposites by 57.5 and 75.6% when embedded with 1 wt% MWNTs and 2 wt% GOs, as well as 2 wt% MWNTs and 4 wt% GOs, respectively. This is because hybrid MWNT/GO reinforcements can achieve much stronger interfacial interactions with PVA matrices leading to higher tensile strengths of their corresponding nanocomposites. On the other hand, Morimune et al. [36] mentioned that with the inclusion of 5 wt% nanodiamonds (NDs), the tensile strength of PVA nanocomposites was enhanced by 30.5%. In comparison, tensile strengths of PVA/NBC bionanocomposites and PVA/MBC bionanocomposites at the BC content of 3 wt% have been reported in this study to be increased by 110 and 72%, which appear to be more desirable than other carbon-based nanocomposites reinforced with GOs and SWNTs (despite relatively small SWNT contents used) when both nanofiller cost and enhancement levels in relation to mechanical properties of nanocomposites are concurrently considered.

3.3.3 Fracture Morphology

As observed in Fig. 3.8, all bionanocomposites have been characterised with rough fracture surfaces due to embedded BC nanoparticles when compared with neat PVA. In particular, uniform multi-layered structures are evidently seen for PVA/NBC bionanocomposites with the inclusion of 3 and 5 wt% NBCs, as illustrated in Fig. 3.8b, c, along the parallel direction to tensile loading. Such structures are very similar to nacre structures mentioned elsewhere [3, 27, 28] and are in good agreement with those observed in PVA-grafted GO (PVA-g-GO) nanocomposites [28]. The fibrillation effect of fractured materials also takes place particularly in PVA/NBC bionanocomposites when subjected to the stretching of bionanocomposite films under the unidirectional tension. This phenomenon suggests far higher ductility of PVA/NBC bionanocomposites when compared with that of PVA/MBC counterparts, which is also proven by their consistently higher elongation at break in Fig. 3.7b.

On the other hand, fracture areas of PVA/MBC bionanocomposites demonstrate less layered structures as opposed to PVA/NBC bionanocomposites at the BC content of 3 wt%. This phenomenon can be ascribed to smaller surface areas of MBCs to reduce interfacial interactions between fillers and PVA matrices. As depicted in Fig. 3.8g, with increasing the MBC content up to 10 wt%, interparticle distance can be significantly reduced owing to large amounts of particle agglomerates. As a result, the formation of local networks of MBC particles to act as stress concentration sites prone to mechanical failure occurs, thus lowering tensile strengths of nanocomposites, as shown in Fig. 3.7a.

3.3.4 Thermal Properties

From our results mentioned earlier, it is believed that effective load transfer from PVA matrices to BCs can be affected by the interaction of strong hydrogen bonding between oxygenated functional groups of BC particles and hydroxyl groups of PVA molecules in terms of the mobility of polymeric chains. DSC results, shown in Fig. 3.9a, b and Table 3.3, indicate that the glass transition temperatures T_g increase

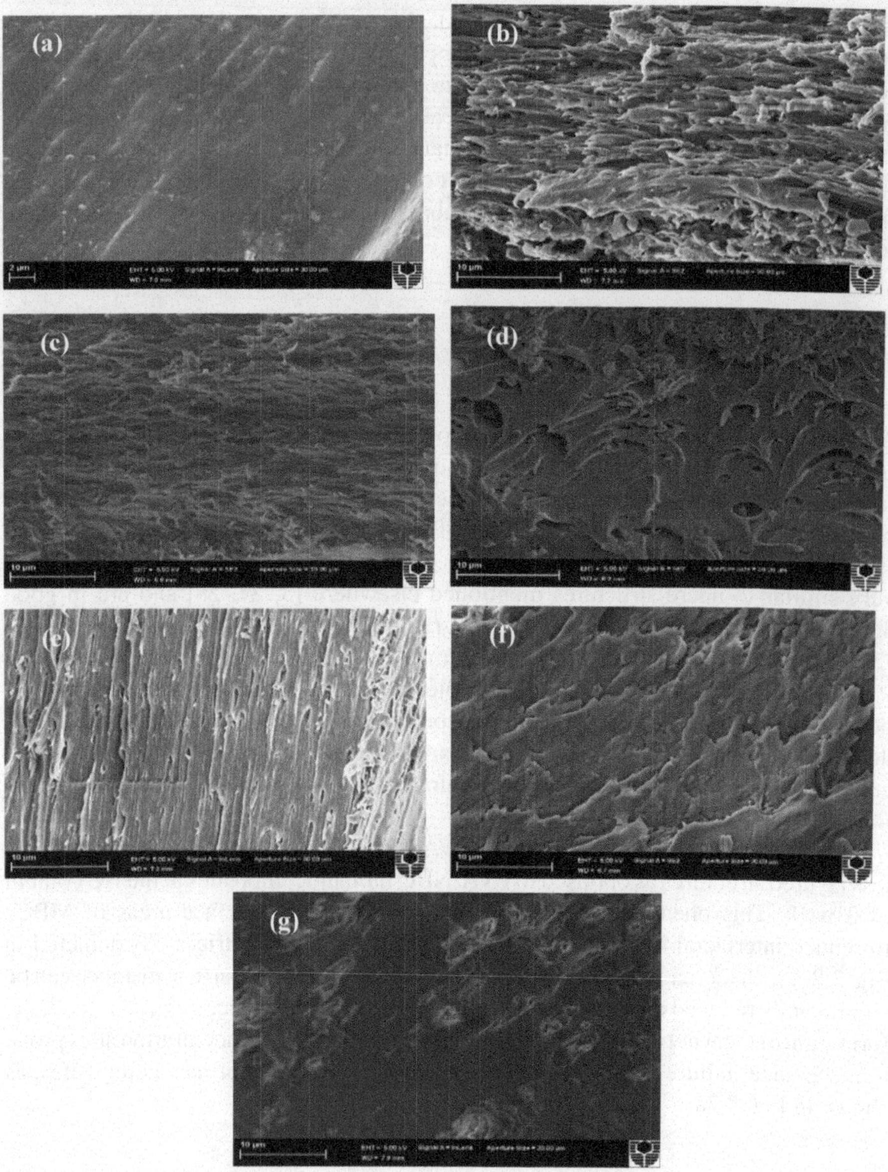

Fig. 3.8 SEM micrographs of different tensile fracture morphologies in **a** PVA, **b** PVA/3 wt% NBC bionanocomposites, **c** PVA/5 wt% NBC bionanocomposites, **d** PVA/10 wt% NBC bionanocomposites, **e** PVA/3 wt% MBC bionanocomposites, **f** PVA/5 wt% MBC bionanocomposites and **g** PVA/10 wt% MBC bionanocomposites [9]

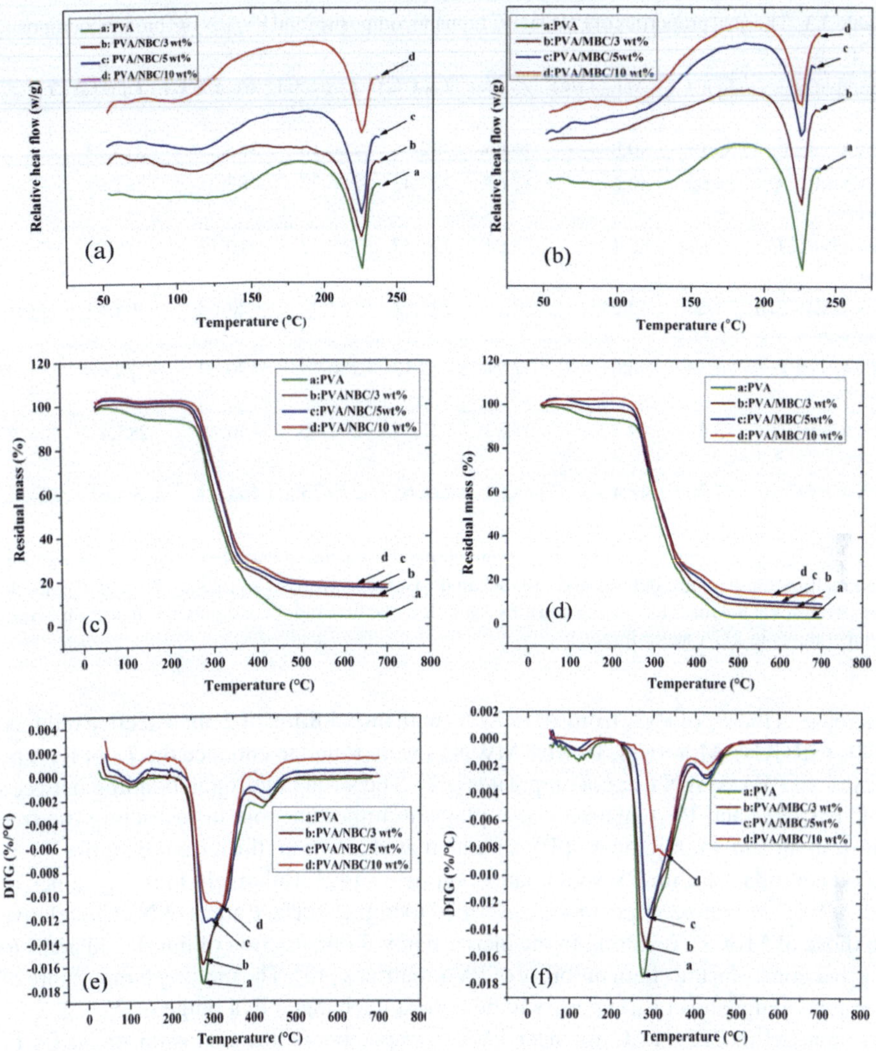

Fig. 3.9 DSC diagrams for **a** PVA/NBC bionanocomposites and **b** PVA/MBC bionanocomposites. TGA curves **c**, **e** as well as DTG curves **d**, **f** for PVA/NBC bionanocomposites and PVA/MBC bionanocomposites, respectively [27, 30]

monotonically up to 75.06 and 73.66 °C for PVA/NBC bionanocomposites and PVA/MBC bionanocomposites, respectively, with increasing the BC content from 0 to 10 wt%, when compared with the T_g of PVA at 65.19 °C. The incorporation of rigid BC particles can restrict the chain mobility of PVA matrices so that higher T_g values are required for the phase change of bionanocomposites from the glassy state to rubbery state. This finding is well known for many types of nanofillers such as nanoclays, GOs, CNTs, HNTs and so on [21]. Manna et al. [40] reported a slight

Table 3.3 Thermal properties of PVA/MBC bionanocomposites and PVA/NBC bionanocomposites [27]

Sample	T_g (°C)	ΔH_m (J/g)	χ_c (%)	T_m (°C)	$T_{5\%}$ (°C)	$T_{80\%}$ (°C)	T_d (°C)	T_d' (°C)
PVA	65.19	50.8	36.65	222.91	200.15	363.5	274.23	
PVA/MBC/3 wt%	68.62	50.6	37.63	223.43	256.22	383.3	281.81	
PVA/MBC/5 wt%	69.89	50.1	38.04	224.37	258.78	410.12	288.93	
PVA/MBC/10 wt%	73.66	49.91	40	224.93	265.9	428.03	305.43	325.03
PVA/NBC/3 wt%	70.53	50.01	37.21	222.12	256.3	390.67	278.48	
PVA/NBC/5 wt%	73.46	49.77	37.83	221.57	262.96	440.28	283.4	303.77
PVA/NBC/10 wt%	75.06	48.16	38.6	222.63	270.73	464.03	294.61	324.12

Note T_g, ΔH_m, χ_c and T_m are glass transition temperature, melting enthalpy, degree of crystallinity and melting temperature determined in DSC analysis, respectively. $T_{5\%}$, $T_{80\%}$, T_d and $T_{d'}$ are the temperatures at the mass losses of 5 and 80%, as well as the first and second maximum decomposition temperatures in TGA accordingly

increase in the T_g of PVA from 70 to 73 °C with the addition of 2 phr selective reduced GOs (SRGOs). Moreover, 10 wt% MWNTs were found to enhance the T_g of PVA by 4.8 °C in PVA/MWNT nanocomposites [41]. The strong hydrogen bonding between BC particles and PVA matrices also plays an important role in restricting the free movement and arrangement of PVA molecular chains with the required higher T_g, in good accordance with PVA/GO nanocomposites [42]. Relatively high T_g values of PVA/NBC bionanocomposites are ascribed to higher surface areas of NBCs (relative to those of MBCs), resulting in the increase in volume fraction of interfacial areas to further constrain the chain mobility of PVA matrices [43]. The melting temperature of bionanocomposites (T_m) appears to be almost unchanged in a range of 221–225 °C as opposed to 222.91 °C for neat PVA, irrespective of the inclusion of NBCs or MBCs. The similar trend also exists that the degree of crystallinity (χ_c) of PVA is only slightly enhanced from 36.65% for neat PVA to 40% and 38.6% for corresponding bionanocomposites with the incorporation of 10 wt% MBCs and NBCs, respectively. This finding suggests that both MBCs and NBCs have relatively minor effect on crystalline phases of PVA matrices, but highly impact their amorphous phases instead [44].

Thermogravimetric analysis (TGA) and derivative thermogravimetric (DTG) curves in terms of mass loss and their derivatives are presented in Fig. 3.9c–f, along with major results listed in Table 3.3. Based on the DTG curve of PVA, there are three major degradation steps according to previous literature [45]. Initially, the first DTG peak at the temperature of 107 °C is associated with the evaporation of absorbed water. Then the second peak at the temperature of 274 °C is attributed to the decomposition of PVA due to the dehydration reaction on polymeric chains, as well as degradation of main backbones. Finally, the third peak in relation to the degradation of polyene residues appears at the temperature below 429 °C. With increasing NBC and MBC contents, the thermal stabilities of both PVA/BC bionanocomposites are improved significantly with reference to that of PVA, as evidenced by consistently higher $T_{5\%}$, $T_{80\%}$ and T_d values. Thermal stability level of bionanocomposites is even more pronounced when incorporated with NBCs in relation to $T_{5\%}$ and $T_{80\%}$. The shift of decomposition temperatures T_d and $T_d^{'}$ means that the dehydration process can be potentially hindered, arising from the interaction between hydroxyl groups of PVA and hydroxyl groups on NBC and MBC surfaces, as confirmed from previous FTIR results. Furthermore, the mass loss process occurring in the third DTG peaks suggests that the thermal decomposition of PVA bionanocomposites requires more reaction activation energy with the higher reaction order [46]. This finding may be attributed to the existence of NBC and MBC particles acting as barrier materials to limit the exothermicity of pyrolysis reaction with better thermal resistance of PVA bionanocomposites. On the other hand, DTG peaks of PVA bionanocomposites at the second decomposition step become much wider than that of neat PVA, consisting of main and side peaks, as opposed to single DTG peak for PVA at the same step. Such a single peak for PVA can be associated with the elimination reaction, while side and main peaks for PVA/BC bionanocomposites correspond to elimination reaction, as well as the overlap of continual eliminations and chain-scission reactions that require more energy to accrue at high temperatures [46].

3.3.5 *Topography and Nanomechanical Properties*

Nanomechanical properties of PVA films are presented in Fig. 3.10. As seen in Fig. 3.10a, there are two typical distinct phases known as crystalline phase (in bright white colour) and amorphous phase (in dark colour), which are generally detected in semicrystalline polymers [47, 48] like PVA. Elastic modulus mapping image of PVA has been analysed in its middle section (A_1-B_1) with associated results being depicted in Fig. 3.10c. The regions with a high modulus level of 24 ± 4.2 GPa and a phase width of 20–76 nm are related to the bundles of semicrystalline stacks in crystalline phase. This value appears to be far higher than the elastic modulus of 9.9 GPa for local PVA/poly (acrylic acid) (PAA) nanophase, as previously reported by Pakzad et al. [49]. Nonetheless, it is very similar to corresponding elastic modulus of 23.69 GPa obtained from PVA/chitosan composite film coating [50]. In comparison, the regions assigned to amorphous phase have a relatively low modulus of 11.4 ± 3.1 GPa along with the amorphus phase width of 18–65 nm. The variation of crystalline phase width is attributed to different orientation of semicrystalline bundles for the designated section [47]. Elastic modulus of bulk PVA films is about 2.08 GPa at a macroscopic level, which is far less than that of local PVA nanoscaled regions based on AFM. Such a phenomenon is manifested when nanomechanical

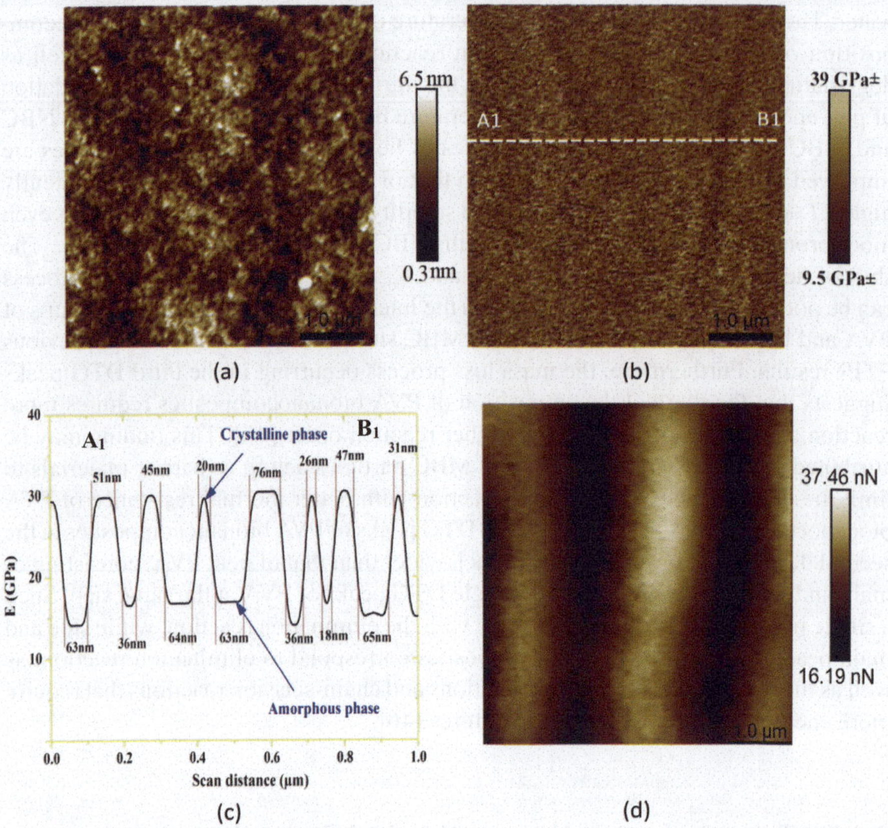

Fig. 3.10 2D height mapping image (**a**) and elastic modulus mapping image (**b**) of PVA films, **c** elastic modulus profile of corresponding section area A_1-B_1 in **b**, **d** 2D adhesion mapping image of PVA films [52]

properties are compared with those of bulk composites [15, 51]. In epoxy/carbon fibre composites, nanomechanical results showed that the modulus of epoxy was 17 GPa as opposed to 3–4 GPa for bulk composites [51]. In a similar manner, local PVA nanoscaled region could yield an elastic modulus of 11.9 GPa in contrast with 2.08 GPa for bulk PVA films [9]. These differences can be related to several reasons. Firstly, though the AFM parameters were calibrated prior to each measurement, the shape function for the topmost tip of the probe could not be accurate enough at low indentation depth [53]. Low indentation depth comes from low indentation force used to induce residual stress and plastic deformation from next indents, thus leading to the maximisation of lateral resolution capabilities, particularly taking place at a nanoscaled level. Secondly, in many cases, local nanosurface properties are different from bulk properties, and the penetration depth strongly influences the measurement of elastic modulus at a nanoscaled level because of the discrepancy occurring in morphological structures between the outer skin on material surfaces and bulk materials [51, 53]. Finally, despite the robust data and good repeatability of calculated elastic moduli, PFQNM method via AFM is inevitably different from conventional tensile testing to determine the tensile modulus for bulk properties due to distinct measurement mechanisms [54]. The difference in PVA phases can also be characterised by the adhesion force mapping image exhibited in Fig. 3.10d, in which adhesion forces for crystalline and amorphous phases have been determined to be 19.7 ± 0.6 and 33 ± 2.8 nN, respectively. The higher adhesion of amorphous phase arises from prevalent viscoelastic material behaviour of amorphous polymers in a non-glassy state. This finding indicates that amorphus phase has less density and surface tension as compared to corresponding crystalline phase [48].

The elastic modulus mapping image of PVA/3 wt% NBC bionanocomposites is assessed on the A_2-B_2 section, as depicted in Fig. 3.11b. The circled peaks with high modulus values can be associated with the presence of NBCs within PVA matrices. In addition, a cyclic pattern of low and high modulus domains depicted in Fig. 3.11b is correlated with the alternating sequence of amorphous and crystalline phases of PVA accordingly, which is considered as a very common characteristic in semicrystalline polymers [47, 48]. Furthermore, the modulus trend of PVA matrices with the addition of 3 wt% NBCs, as shown in Fig. 3.11b, is totally different from that of neat PVA along with the change to the degree of structural orderness, Fig. 3.11c. Such a difference may be attributed to reinforcing effect of NBCs on both crystalline and amorphous phases of PVA matrices. The morphological structures of bionanocomposites in term of elastic modulus have been further investigated at a high magnification in Fig. 3.11c. Furthermore, a data analysis in the middle of the mapping image (A_3-B_3) is depicted in Fig. 3.10d. It is worth noting that both crystalline and amorphous phase widths are in a range of 5–53 and 4–35 nm, respectively, which, however, are lower than that of corresponding PVA (Fig. 3.10c). Decreasing the phase width means that the stack size decreases accordingly from neat polymer to bionanocomposite films. In other words, the number of lamellae stacks per unit volume, on the other hand, increases from neat polymer to bionanocomposites illustrated in Fig. 3.11e, thus resulting in the improvement of tensile strengths of PVA bionanocomposites. This phenomenon can be interpreted by the Hall–Petch relation [55, 56] where yield stress and tensile stress depend on the grain size. As a consequence, much higher tensile strength of PVA/NBC nanocomposites at 147.94 MPa can be achieved as opposed to that of PVA at 70.32 MPa. These findings offer the insight to establish a more accurate theoretical modelling framework by combining both nanomechanical properties of bionanocomposites and their bulk properties at the macroscopic level.

In case of PVA/3 wt% MBC bionanocomposites, the mapping image of elastic modulus is shown in Fig. 3.12a, and the data analysis in the mapping image (A_4-B_4) is depicted in Fig. 3.12b. The crystalline phase width is in a range of 14–60 nm when compared with the amorphous width of 15–59 nm, which is still lower than that of PVA. This finding suggests that the number of lamellae stacks per unit volume increases with the inclusion of MBCs as compared with that of pure PVA. Such a result also explains why higher tensile strength of PVA/3 wt% MBC bionanocomposites can be achieved as opposed to that of pure PVA.

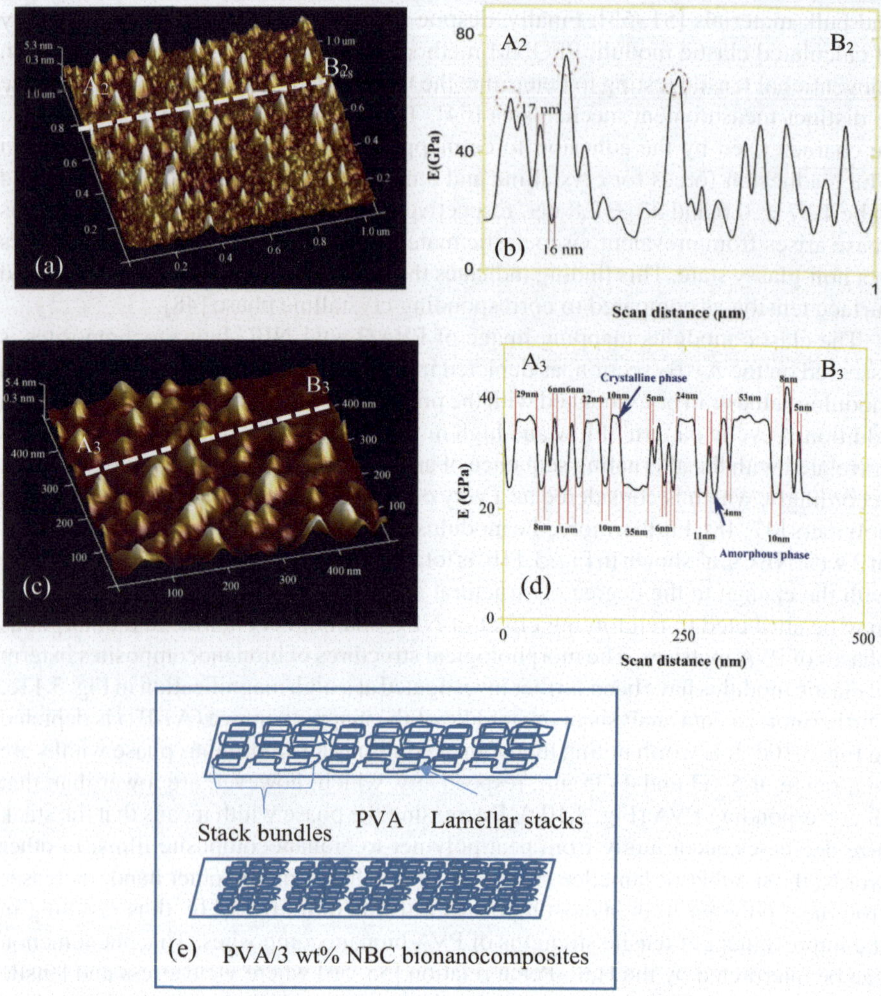

Fig. 3.11 **a** 3D height mapping image of PVA/3wt% NBC bionanocomposites and **b** elastic modulus profile curve for corresponding section area A_2-B_2 in **a**, **c** 3D height mapping image of PVA/3wt% NBC bionanocomposites at a high magnification, **d** elastic modulus profile for corresponding section area A_3-B_3 in **c**, as well as **e** schematic diagram for bundles of lamellar stacks in both neat PVA and PVA/3 wt% NBC bionanocomposites [52]

However, in comparison with PVA/3 wt% NBC bionanocomposites, MBC inclusion shows less apparent effect on crystalline and amorphous phases of PVA. In addition, the widths of amorphous and crystalline phases are still higher than those of PVA/3 wt% NBC bionanocomposites. Such a finding can be related to much higher surface areas of NBCs, which makes them cover more contact areas within PVA matrices, and then increases the volume fraction of interphase, as compared with that of MBCs.

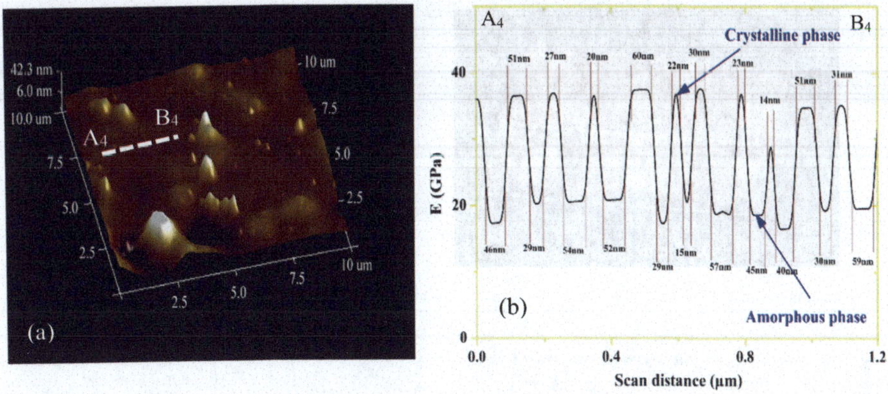

Fig. 3.12 **a** 3D height mapping image of PVA/3 wt% MBC bionanocomposites and **b** elastic modulus profile curve for corresponding section area A4-B4 shown in **a**

Figure 3.13 demonstrates 3D height mapping images and corresponding section profiles of PVA and PVA/NBC bionanocomposites. When the NBC content is below 10 wt%, spiky NBCs appear to be separated from one another resulting in homogeneous NBC dispersion within PVA matrices, Fig. 3.13b, c. Besides, average root-mean-square (R_q) values for neat PVA and PVA/NBC bionanocomposite films have been reported with the focus of 1 μm $\times$ 1 μm surface area in Fig. 3.13. When the NBC content increases from 3 to 5 wt%, R_q values of PVA/NBC bionanocomposites are enhanced from 2.1 $\pm$ 0.11 to 2.5 $\pm$ 0.16 nm, which are very close to 1.9 $\pm$ 0.17 nm for neat PVA. This finding suggests that smooth PVA/NBC bionanocomposite film surfaces are manifested at low NBC contents as expected, which is in good accordance with the experimental result of PVA/nanocellulose composite films [57]. On the contrary, the inclusion of 10 wt% NBCs in bionanocomposites leads to a nearly two-fold R_q value of 4.1 $\pm$ 0.19 nm as opposed to the counterparts with the inclusion of 3 wt% NBCs, which clearly signifies that the presence of aggregated NBCs results in much higher surface roughness on PVA surfaces [58]. Nonetheless, this value of 4.1 $\pm$ 0.19 nm remains relatively low when compared with other PVA nanocomposites reinforced with carbon-based fillers such as PVA/reduction GO (rGO) nanocomposites with a R_q value of 4.6 $\pm$ 0.55 nm based on deposition layers [59].

In case of PVA/MBC bionanocomposite films, relevant results reveal that MBC particles are dispersed in a uniform manner when the MBC content increases up to 5 wt%, as depicted in Fig. 3.14a, b. However, an excessive amount of MBCs causes a decrease in intraparticle spacing, and MBC agglomeration would inevitably take place, as illustrated in Fig. 3.14c. Moreover, the R_q values for PVA/MBC bionanocomposites at the MBC contents of 3, 5 and 10 wt% have been determined to be 5.47 $\pm$ 1.23, 8.36 $\pm$ 1.98 and 13.7 $\pm$ 2.12 nm, respectively, which appear to be higher than those of pure PVA and PVA/NBC bionanocomposites. Such results

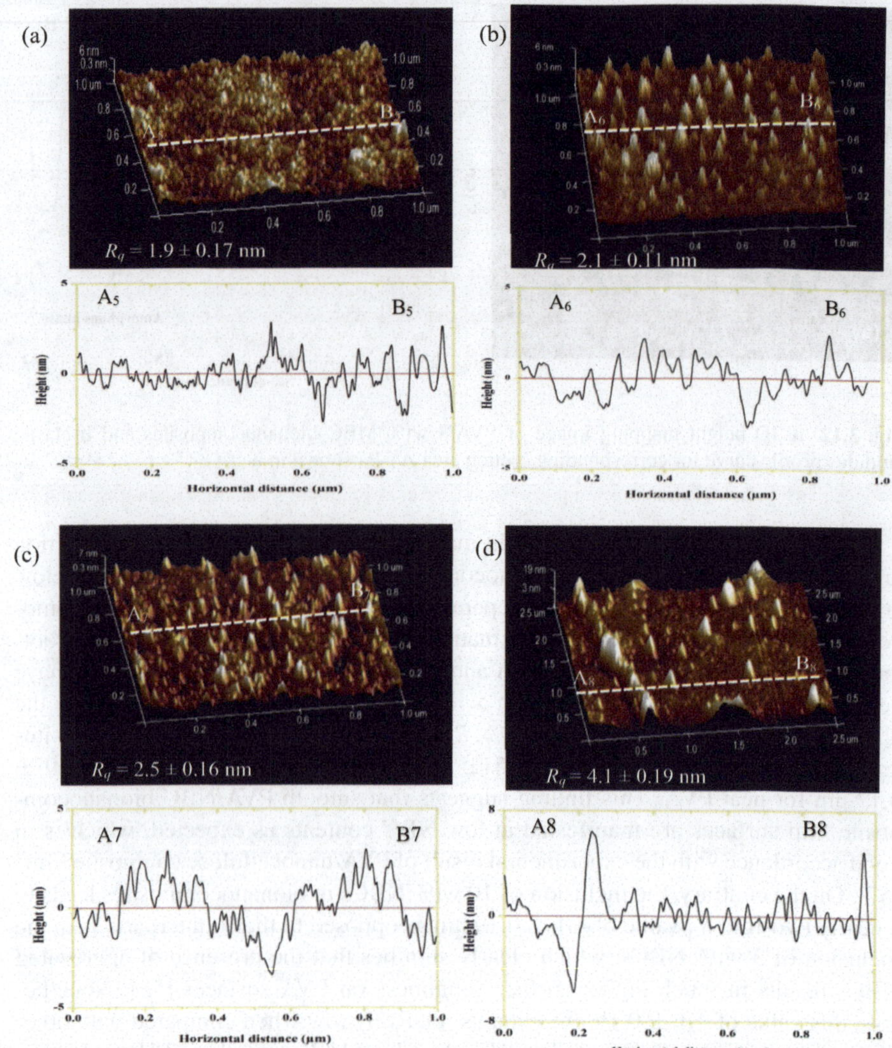

Fig. 3.13 3D height mapping images and corresponding height profiles for **a** PVA, **b** PVA/3 wt% NBC bionanocomposites, **c** PVA/5 wt% NBC bionanocomposites and **d** PVA/10 wt% NBC bionanocomposites [30]

are attributed to the larger size and greater thickness of MBCs when compared with those of NBCs.

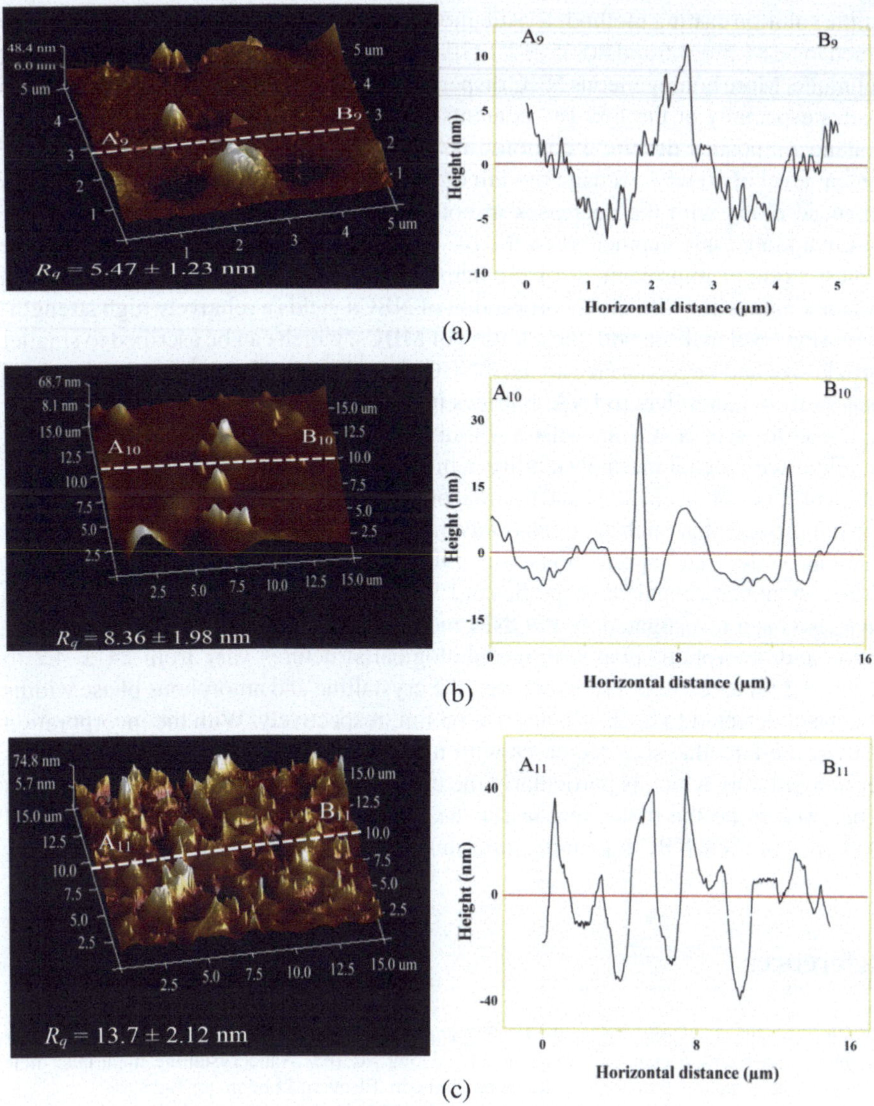

Fig. 3.14 3D height mapping images and corresponding height profiles for **a** PVA/3 wt% MBC bionanocomposites; **b** PVA/5 wt% MBC bionanocomposites and **c** PVA/10 wt% MBC bionanocomposites

3.4 Summary

This chapter covers the manufacture and material characterisation of strong and eco-friendly PVA/NBC bionanocomposites and PVA/MBC bionanocomposites using a simple solution casting method. Elastic moduli of NBCs and MBCs have been determined to be 84.5 ± 3.6 and 80.65 ± 2.1 GPa for the first time via a sophisticated AFM technique. More homogeneous NBC dispersion is shown for PVA/NBC bionanocomposites especially at the low BC contents up to 3 wt%, as opposed to PVA/MBC bionanocomposites despite a common issue of BC agglomeration at the high BC content level of 10 wt%. Tensile moduli of both bionanocomposites are significantly enhanced along with the decreases in both elongation at break and tensile toughness in a monotonic manner when the BC content increases from 0 to 10 wt%. The highest strength improvement of bionanocomposites appears at 3 wt%, and such bionanocomposites with the incorporation of NBCs yields a relatively high strength-increasing level to those with the addition of MBCs, which can be ascribed to smaller particle size and larger surface areas of NBCs in order to achieve more effective load transfer from nanofillers to PVA matrices in PVA/BC bionanocomposites.

The addition of NBCs and MBCs within PVA matrices also reveals the increase of T_g values with higher thermal stability in bionanocomposite films. With the determination of superior mechanical and thermal properties of PVA/BC bionanocomposites, it is anticipated that such fabricated advanced composite materials can be utilised for more widespread applications with full biodegradability and biocompatibility. Moreover, nanomechanical properties of PVA and PVA/BC bionanocomposite films have also been investigated by PFQNM measurements. The elastic moduli of crystalline and amorphous phases in morphological structures vary from 24 ± 4.2 to 11.4 ± 3.1 GPa, respectively. Moreover, the crystalline and amorphous phase widths have been detected to be 20–76 and 18–65 nm, respectively. With the incorporation of BCs, the lamellae size decreases with increasing the number of lamellae stacks per unit volume, which is particularly the case when incorporated with NBCs. This is believed to be the major reason for the significant improvement of mechanical properties of PVA/NBC bionanocomposites.

References

1. Zhu J, Jia J, Tjong SC (2014) Preparation, structure, and application of carbon nanotubes/bamboo charcoal composite. In: Tjong SC (ed) Nanocrystalline materials: their synthesis-structure-property relationships, 2nd edn. Elsevier, London, pp 1–25
2. Gray M, Johnson MG, Dragila MI, Kleber M (2014) Water uptake in biochars: the roles of porosity and hydrophobicity. Biomass Bioenerg 61:196–205
3. Brockhoff SR, Christians NE, Killorn RJ, Horton R, Davis DD (2010) Physical and mineral-nutrition properties of sand-based turfgrass root zones amended with biochar. Agron J 102(6):1627–1631
4. Ho MP, Lau KT, Wang H, Hui D (2015) Improvement on the properties of polylactic acid (PLA) using bamboo charcoal particles. Compos Part B Eng 81:14–25
5. You Z, Li D (2014) Highly filled bamboo charcoal powder reinforced ultra-high molecular weight polyethylene. Mater Lett 122:121–124
6. Wu KH, Ting TH, Wang GP, Yang CC, Tsai CW (2008) Synthesis and microwave electromagnetic characteristics of bamboo charcoal/polyaniline composites in 2–40 GHz. Synth Met 158(17–18):688–694

7. Li S, Li X, Chen C, Wang H, Deng Q, Gong M, Li D (2016) Development of electrically conductive nano bamboo charcoal/ultra-high molecular weight polyethylene composites with a segregated network. Compos Sci Technol 132:31–37
8. Srinivasan P, Sarmah AK, Smernik R, Das O, Farid M, Gao W (2015) A feasibility study of agricultural and sewage biomass as biochar, bioenergy and biocomposite feedstock: production, characterization and potential applications. Sci Total Environ 512:495–5053
9. Mousa M, Dong Y (2017) Strong poly (vinyl alcohol)(PVA)/bamboo charcoal (BC) nanocomposite films with particle size effect. ACS Sustain Chem Eng 6(1):467–479
10. Das O, Sarmah AK, Bhattacharyya D (2015) A novel approach in organic waste utilization through biochar addition in wood/polypropylene composites. Waste Manage 38:132–140
11. She B, Tao X, Huang T, Lu G, Zhou Z, Guo C, Dang Z (2016) Effects of nano bamboo charcoal on PAHs-degrading strain *Sphingomonas* sp. GY2B. Ecotoxicol Environ Saf 125:35–42
12. Wang XJ, Wang Y, Wang X, Liu M, Xia SQ, Yin DQ, Zhang YL, Zhao JF (2011) Microwave-assisted preparation of bamboo charcoal-based iron-containing adsorbents for Cr (VI) removal. Chem Eng J 174(1):326–332
13. Pirzada T, Arvidson SA, Saquing CD, Shah SS, Khan SA (2012) Hybrid silica–PVA nanofibers via sol–gel electrospinning. Langmuir 28(13):5834–5844
14. Yang X, Li L, Shang S, Tao XM (2010) Synthesis and characterization of layer-aligned poly (vinyl alcohol)/graphene nanocomposites. Polymer 51(15):3431–3435
15. Niu YF, Yang Y, Gao S, Yao JW (2016) Mechanical mapping of the interphase in carbon fiber reinforced poly (ether-ether-ketone) composites using peak force atomic force microscopy: interphase shrinkage under coupled ultraviolet and hydro-thermal exposure. Polym Testing 55:257–260
16. Hossain MA, Islam S, Chowdhury FA, Mohiuddin TG, Uchida K, Tamura T, Sugawa K, Mochida T, Otsuki J, Alam MS (2016) Structural, mechanical, and electrical properties of carbon nanoparticles synthesized from diesel. Fuller Nanotub Car N 24(1):43–51
17. Suk JW, Piner RD, An J, Ruoff RS (2010) Mechanical properties of monolayer graphene oxide. ACS Nano 4(11):6557–6564
18. Demczyk BG, Wang YM, Cumings J, Hetman M, Han W, Zettl A, Ritchie RO (2002) Direct mechanical measurement of the tensile strength and elastic modulus of multiwalled carbon nanotubes. Mater Sci Eng, A 334(1–2):173–178
19. Putz KW, Compton OC, Palmeri MJ, Nguyen ST, Brinson LC (2010) High-nanofiller-content graphene oxide–polymer nanocomposites via vacuum-assisted self-assembly. Adv Funct Mater 20(19):3322–3329
20. Chang Y, Yang ST, Liu JH, Dong E, Wang Y, Cao A, Liu Y, Wang H (2011) In vitro toxicity evaluation of graphene oxide on A549 cells. Toxicol Lett 200(3):201–210
21. Mousa MH, Dong Y, Davies IJ (2016) Recent advances in bionanocomposites: preparation, properties, and applications. Int J Polym Mater Polym Biomater 65(5):225–254
22. Fu D, Zhang Y, Lv F, Chu PK, Shang J (2012) Removal of organic materials from TNT red water by bamboo charcoal adsorption. Chem Eng J 193:39–49
23. Lorenzon M, Evangelio L, Verhaeghe S, Nicolet C, Navarro C, Pérez-Murano F (2015) Assessing the local nanomechanical properties of self-assembled block copolymer thin films by peak force tapping. Langmuir 31(42):11630–11638
24. Li S, Li X, Deng Q, Li D (2015) Three kinds of charcoal powder reinforced ultra-high molecular weight polyethylene composites with excellent mechanical and electrical properties. Mater Des 85:54–59
25. Yang E, Yao C, Liu Y, Zhang C, Jia L, Li D, Fu Z, Sun D, Kirk SR, Yin D (2018) Bamboo-derived porous biochar for efficient adsorption removal of dibenzothiophene from model fuel. Fuel 211:121–129
26. Lu L, Sun H, Peng F, Jiang Z (2006) Novel graphite-filled PVA/CS hybrid membrane for pervaporation of benzene/cyclohexane mixtures. J Membrane Sci 281(1–2):245–252
27. Mousa MH (2018) Experimental characterisation and modelling of sustainable multiscaled bionanocomposites. Ph.D. thesis, Curtin University, Perth, Australia

28. Cheng HKF, Sahoo NG, Tan YP, Pan Y, Bao H, Li L, Chan SH, Zhao J (2012) Poly (vinyl alcohol) nanocomposites filled with poly (vinyl alcohol)-grafted graphene oxide. ACS Appl Mater Interfaces 4(5):2387–2394
29. Zhang J, Lei W, Liu D, Wang X (2017) Synergistic influence from the hybridization of boron nitride and graphene oxide nanosheets on the thermal conductivity and mechanical properties of polymer nanocomposites. Compos Sci Technol 151:252–257
30. Mousa M, Dong Y (2020) The role of nanoparticle shapes and structures in the material characterisation of polyvinyl alcohol (PVA) bionanocomposite films. Polymers 12:264
31. Liang J, Huang Y, Zhang L, Wang Y, Ma Y, Guo T, Chen Y (2009) Molecular-level dispersion of graphene into poly (vinyl alcohol) and effective reinforcement of their nanocomposites. Adv Funct Mater 19(14):2297–2302
32. Xu Y, Hong W, Bai H, Li C, Shi G (2009) Strong and ductile poly (vinyl alcohol)/graphene oxide composite films with a layered structure. Carbon 47(15):3538–3543
33. Arao Y, Mizuno Y, Araki K, Kubouchi M (2016) Mass production of high-aspect-ratio few-layer-graphene by high-speed laminar flow. Carbon 102:330–338
34. Liu L, Barber AH, Nuriel S, Wagner HD (2005) Mechanical properties of functionalized single-walled carbon-nanotube/poly (vinyl alcohol) nanocomposites. Adv Funct Mater 15(6):975–980
35. Li Y, Yang T, Yu T, Zheng L, Liao K (2011) Synergistic effect of hybrid carbon nantube–graphene oxide as a nanofiller in enhancing the mechanical properties of PVA composites. J Mater Chem 21(29):10844–10851
36. Morimune S, Kotera M, Nishino T, Goto K, Hata K (2011) Poly (vinyl alcohol) nanocomposites with nanodiamond. Macromolecules 44(11):4415–4421
37. Paiva MC, Zhou B, Fernando KAS, Lin Y, Kennedy JM, Sun YP (2004) Mechanical and morphological characterization of polymer–carbon nanocomposites from functionalized carbon nanotubes. Carbon 42(14):2849–2854
38. Wang Y, Shi Z, Yin J (2010) Unzipped multiwalled carbon nanotubes for mechanical reinforcement of polymer composites. J Phys Chem C 114(46):19621–19628
39. Chen W, Tao X, Xue P, Cheng X (2005) Enhanced mechanical properties and morphological characterizations of poly (vinyl alcohol)–carbon nanotube composite films. Appl Surf Sci 252(5):1404–1409
40. Manna K, Srivastava SK, Mittal V (2016) Role of enhanced hydrogen bonding of selectively reduced graphite oxide in fabrication of poly (vinyl alcohol) nanocomposites in water as EMI shielding material. J Phys Chem C 120(30):17011–17023
41. Du FP, Ye EZ, Yang W, Shen TH, Tang CY, Xie XL, Zhou XP, Law WC (2015) Electroactive shape memory polymer based on optimized multi-walled carbon nanotubes/polyvinyl alcohol nanocomposites. Compost B Eng 68:170–175
42. Salavagione HJ, Gomez MA, Martinez G (2009) Polymeric modification of graphene through esterification of graphite oxide and poly (vinyl alcohol). Macromolecules 42(17):6331–6334
43. Miaudet P, Derré A, Maugey M, Zakri C, Piccione PM, Inoubli R, Poulin P (2007) Shape and temperature memory of nanocomposites with broadened glass transition. Science 318(5854):1294–1296
44. Leitão A, Silva J, Dourado F, Gama M (2013) Production and characterization of a new bacterial cellulose/poly (vinyl alcohol) nanocomposite. Materials 6(5):1956–1966
45. Qiao R, Deng H, Putz KW, Brinson LC (2011) Effect of particle agglomeration and interphase on the glass transition temperature of polymer nanocomposites. J Polym Sci, Part B: Polym Phys 49(10):740–748
46. Peng Z, Kong LX (2007) A thermal degradation mechanism of polyvinyl alcohol/silica nanocomposites. Polym Degrad Stab 92(6):1061–1071
47. Panaitescu DM, Frone AN, Nicolae C (2013) Micro-and nano-mechanical characterization of polyamide 11 and its composites containing cellulose nanofibers. Eur Polym J 49(12):3857–3866
48. Voss A, Stark RW, Dietz C (2014) Surface versus volume properties on the nanoscale: elastomeric polypropylene. Macromolecules 47(15):5236–5245

49. Pakzad A, Simonsen J, Yassar RS (2012) Gradient of nanomechanical properties in the interphase of cellulose nanocrystal composites. Compos Sci Technol 72(2):314–319
50. Mishra SK, Kannan S (2014) Development, mechanical evaluation and surface characteristics of chitosan/polyvinyl alcohol based polymer composite coatings on titanium metal. J Mech Behav Biomed Mater 40:314–324
51. Gu Y, Li M, Wang J, Zhang Z (2010) Characterization of the interphase in carbon fiber/polymer composites using a nanoscale dynamic mechanical imaging technique. Carbon 48(11):3229–3235
52. Mousa M, Dong Y (2018) Elastic behavior of nanophases in polyvinyl alcohol (PVA)/bamboo charcoal (BC) nanocomposite films. Front Mater 5:44
53. Clifford CA, Seah MP (2005) Quantification issues in the identification of nanoscale regions of homopolymers using modulus measurement via AFM nanoindentation. Appl Surf Sci 252(5):1915–1933
54. Syed Asif SA, Wahl KJ, Colton RJ, Warren OL (2001) Quantitative imaging of nanoscale mechanical properties using hybrid nanoindentation and force modulation. J Appl Phys 90(3):1192–1200
55. Hall EO (1951) The deformation and ageing of mild steel: III discussion of results. Proc Phys Soc B 64(9):747
56. Song K, Zhang Y, Minus ML (2015) Polymer interphase self-reinforcement and strengthening mechanisms in low-loaded nanocomposite fibers. Macromol Chem Phys 216(12):1313–1320
57. Hakalahti M, Mautner A, Johansson LS, Hänninen T, Setälä H, Kontturi E, Bismarck A, Tammelin T (2016) Direct interfacial modification of nanocellulose films for thermoresponsive membrane templates. ACS Appl Mater Interfaces 8(5):2923–2927
58. Liu M, Zhou C, Dong B, Wu Z, Wang L, Yu S, Gao C (2014) Enhancing the permselectivity of thin-film composite poly (vinyl alcohol)(PVA) nanofiltration membrane by incorporating poly (sodium-p-styrene-sulfonate)(PSSNa). J Membrane Sci 463:173–182
59. Humood M, Qin S, Song Y, Polychronopoulou K, Zhang Y, Grunlan JC, Polycarpou AA (2016) Influence of graphene reduction and polymer cross-linking on improving the interfacial properties of multilayer thin films. ACS Appl Mater Interfaces 9(1):1107–1118

Chapter 4
PVA Bionanocomposite Films with Different Particle Shapes and Structures

Abstract In this chapter, three different types of nanoparticles, namely 1D Cloisite 30B MMT clay nanoplatelets, 2D halloysite nanotubes (HNTs) and 3D nanodiameter bamboo charcoals (NBCs), were employed to investigate the impact of nanoparticle shapes and structures on the material performance of polyvinyl alcohol (PVA) bionanocomposite films from the perspectives of mechanical and thermal properties, morphological structures and nanomechanical behaviour. The overall results revealed superior reinforcement efficiency of NBCs to both Cloisite 30B clays and HNTs owing to their typical porous structures to actively interact with PVA matrices in the combination of strong mechanical and hydrogen bondings. 3D NBCs also achieved better nanoparticle dispersibility when compared with 1D Cloisite 30B clays and 2D HNTs along with higher thermal stability, which was attributed to larger interfacial regions when characterised for the nanomechanical behaviour of corresponding bionanocomposite films. This chapter offers an insightful guidance to an appropriate selection of nanoparticles as effective reinforcements and the further sophisticated design of bionanocomposite materials.

Keywords Polyvinyl alcohol (PVA) · Halloysite nanotubes (HNTs) · Clay nanoplatelets · Bamboo charcoals (BCs) · Bionanocomposites · Nanostructures · Material characterisation

4.1 Introduction

Nanoparticles in spheroidal, platelet-like and tubular shapes as effective nanofillers have drawn great attention to materials engineers and researchers in the field of nanocomposite materials over last few decades [1, 2]. The incorporation of those different types of nanoparticles into continuous polymer matrices has been proven to significantly alter the properties of base polymers, resulting in a novel-class system of polymer nanocomposites with superior properties and excellent functionalities [1, 2]. In general, when incorporated with a small fraction of nanoparticles being less than

This chapter takes the most of content materials from authors' published research article **'The role of nanoparticle shape and structures in materials characterisation of polyvinyl alcohol (PVA) bionanocomposite films' Polymers 12 (2020)264** under Creative Commons Attribution License.

M. Mousa and Y. Dong, *Multiscaled PVA Bionanocomposite Films*,
https://doi.org/10.1007/978-981-15-8771-9_4

10 wt%, optical [3], mechanical [4], thermal [5], electronic [6] and antimicrobial [7] properties can be remarkably enhanced in polymer nanocomposites, while some key features of net polymer systems remain such as low density and high processibility [2]. Such polymer nanocomposites have a wide range of applications in range of aerospace engineering, automotive components and medical devices [1]. The effective reinforcing mechanism is based on a fundamental concept that the chain mobility of polymeric molecules is restricted by rigid nanofillers according to matrix–particle interfacial interactions in polymer nanocomposites [2, 8]. The specific areas associated with matrix–filler interactions are known as interfacial regions with completely distinct properties from nanoparticles and polymer matrices alone. More importantly, the material performance of polymer nanocomposites primarily depends on the volume of interfacial regions and properties [8] in relation to critical nanofiller factors such as nanoparticle shapes and structures.

In addition to a major concern of nanoparticle structures, nanoparticle shapes are also equally important when matrix–filler interaction is considered in polymer nanocomposites, which can be classified into three popular shapes, namely 1D platelet-like nanoparticles such as montmorillonite (MMT) clays and graphene nanoplatelet sheets, 2D tubular nanoparticles like HNTs and carbon nanotubes (CNTs) and 3D spherical or near-spherical nanoparticles including NBCs, diamond nanoparticles and nanosilica particles. In a nanocomposite system, the alteration of nanoparticle shapes means that the contact areas inevitably vary between polymer matrices and nanoparticles to effectively control the volume of interfacial regions between them [8]. Most previous studies [9, 10] were based on theoretical or numerical modelling approaches like atomistic and coarse-grained molecular dynamic (MD) simulations for evaluating matrix–filler bonding effect. Nonetheless, current computational capability and environment may be mostly restricted to the context of single- and two-particle systems by neglecting the effect of actual nanoparticle structures and shapes that are induced in different material processing techniques [11].

4.2 Nanoparticle Shape and Size

Morphological structures of as-received nanoparticles of Cloisite 30B clays, HNTs and NBCs are illustrated in Fig. 4.1. All nanoparticle powders show high irregularities in size and material morphology. However, HNTs are most likely to possess cylindrical shapes with transparent central areas running longitudinally along such cylindrical structures, as illustrated in Fig. 4.1a, b. The outer diameters and lumen diameters of HNTs, as typical tubular nanoparticles in hollow and open-end structures, were found to be in a range of 20–115 and 5–30 nm, respectively, while the lengths of HNTs vary from 50 nm to 1.5 μm. On the other hand, morphological structures of Cloisite 30B clays were detected using an AFM tapping mode from diluted clay suspension when deposited onto the mica substrate, as revealed in Fig. 4.1c, d. It is evidently seen that Cloisite 30B clays possess platelet-like structures with average

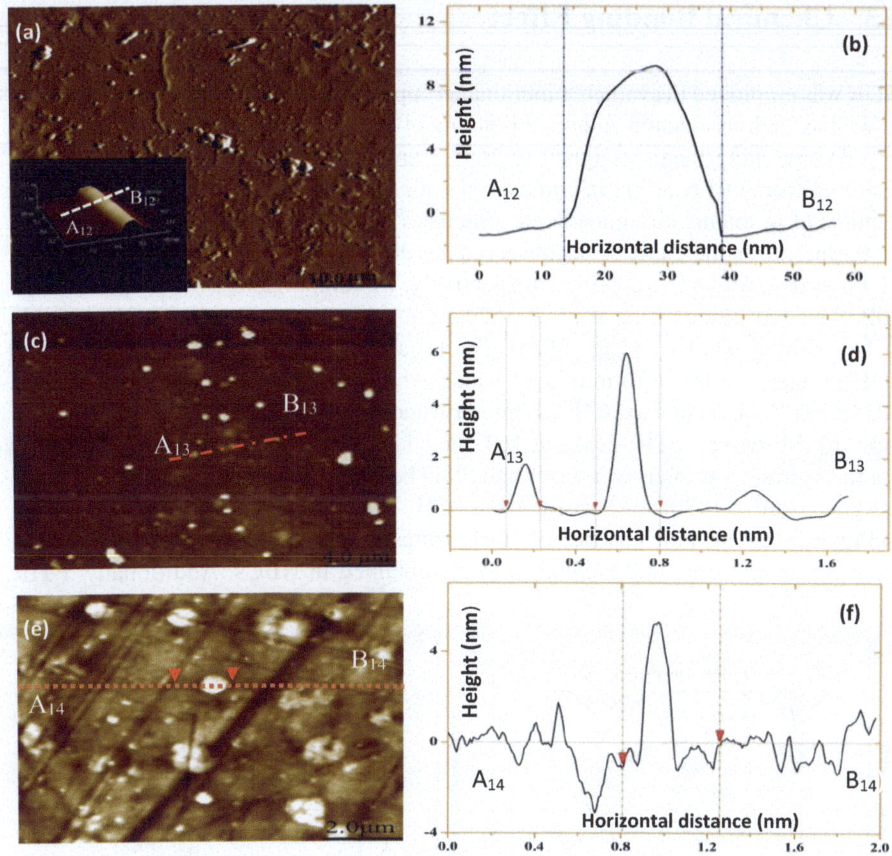

Fig. 4.1 AFM characterisation of different nanoparticles: AFM images of **a** HNTs, **c** Cloisite 30B clays and **e** NBCs deposited on mica substrate in aqueous solutions, and height profiles of **b** HNTs, **d** Cloisite 30B clays and **f** NBCs at typical cut sections A_{12}-B_{12}, A_{13}-B_{13} and A_{14}-B_{14}, respectively [12]

particle diameter of approximately 100.75 ± 6.5 nm according to 925 measured clay particles, while their thickness varies from 1.69 to 5.9 nm, which is exemplified by cross-sectional analysis for a typical cut section (A_{13}-B_{13}) illustrated in Fig. 4.1d. These results suggest that clay platelet-like layered structures can consist of single platelet with the thickness of nearly 1 nm [13], as well as stacked layered platelets, apparently indicating the combined effect of clay exfoliation and intercalation. The size of NBCs was determined previously to be 69.43 nm in average size [4], as shown in Fig. 4.1e, f, which can be considered as 3D anisotropic nanoparticles when compared with 1D platelet-like Cloisite 30B clays and 2D HNTs.

4.3 Chemical Bonding Effect

FTIR was employed to evaluate functional groups in PVA matrices and nanoparticles, as well as their associated chemical bonding effects. As-received HNTs exhibit two Al_2OH stretching bands assigned to 3691.5 and 3621 cm^{-1} depicted in Fig. 4.2a, resulting from OH bending in connection with two Al atoms along with other band features of inorganic aluminosilicate structures of halloysite [14].

Furthermore, the FTIR peaks observed at 1004 and 906 cm^{-1} are associated with Si–O–Si and Al–OH stretchings, respectively. In comparison, as-received Cloisite 30B clays have an existing peak at 3629.6 cm^{-1} corresponding to Si–OH and Al–OH stretchings shown in Fig. 4.2(b). The broad band at 3405 cm^{-1} is assigned to –OH groups in relation to interlayer water, while two typical bands at 2924.6 and 2853.5 cm^{-1} are related to $-CH_2$ asymmetric and symmetric stretchings, respectively [15, 16]. Moreover, FTIR peaks at 1647 and 1123.3 cm^{-1} can be due to the deformation vibration of interlayer water and Si–O bending accordingly, as opposed to an assigned band at 1470 cm^{-1} arising from $-CH_2$ bending [15]. NBC results presented in Fig. 4.2c confirm the absence of –OH groups in their FTIR spectra, which infers much lower moisture and alcohol contents obtained in NBCs. Additionally, FTIR

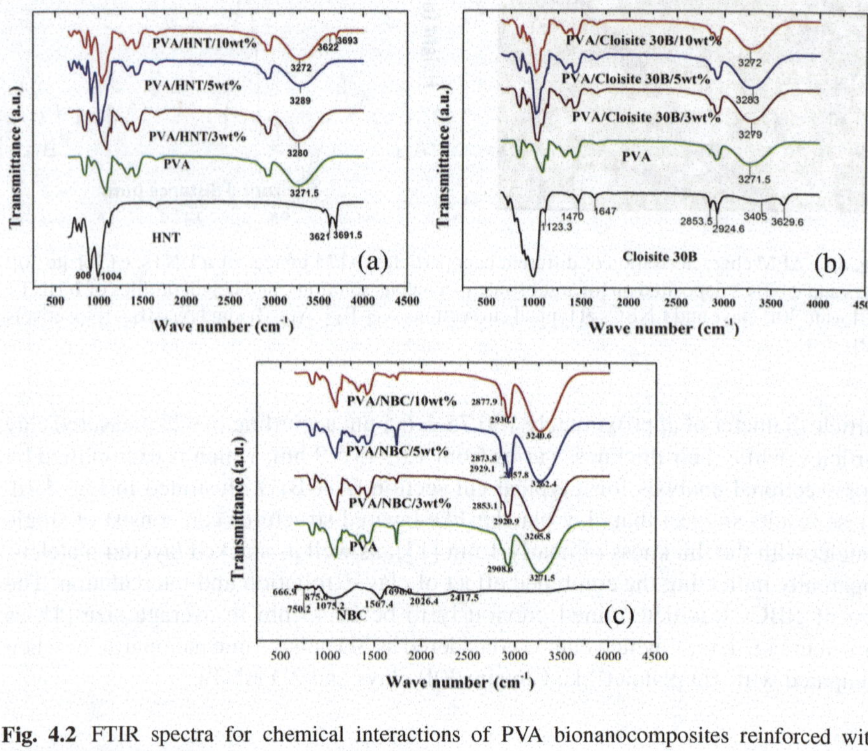

Fig. 4.2 FTIR spectra for chemical interactions of PVA bionanocomposites reinforced with **a** HNTs, **b** Cloisite 30B clays and **c** NBCs [12]

peaks detected at 2417.5 and 1567.4 cm^{-1} reveal the existence of C≡H stretching [17] and C=C vibration in an aromatic system [18], respectively. On the other hand, the peak spectrum at 1696 cm^{-1} is assigned to C=O band primarily for ionisable carboxyl groups as an indicator of surface hydrophilicity [19]. In case of bionanocomposite systems, the FTIR spectra of PVA/HNT bionanocomposites and PVA/Cloisite 30B clay bionanocomposites are also illustrated in Fig. 4.2a, b, respectively. The FTIR peak located at 3271.5 cm^{-1} is associated with O–H stretching, which shifts to higher wavenumbers at 3280 and 3289 cm^{-1}, as well as 3279 and 3283 cm^{-1} with the inclusion of HNTs and Cloisite 30B clays at the nanoparticle contents of 3 and 5 wt%, respectively. Such a finding is attributed to the strengthening effect of hydrogen bonds between –OH groups obtained from PVA molecules and those located on Cloisite 30B clay surfaces such as silanol groups (–SiOH), which is in good agreement with previous investigations on PVA/organomodified Cloisite Na^{+} (OMMT) nanocomposites [16], poly (ε-caprolactone) (PCL)/Cloisite 30B clay nanocomposites [20], as well as PVA/chitosan (CS)/HNT nanocomposites [21]. However, when the HNT content increases up to 10 wt%, two Al_2OH stretchings appear with respect to embedded HNTs in bionanocomposite films, which can be ascribed to typical HNT agglomeration [22]. As for PVA/NBC bionanocomposite films, increasing the NBC content from 0 to 10 wt% yields the band-peak shift to lower wavenumbers at 3240.6 and 3245.5 cm^{-1}, respectively, due to large amounts of hydroxyl groups in PVA molecules [4], as well as strong filler–matrix interactions. As a result, hydrogen bonds are generated to be intertwined at PVA/NBC interfaces with a broad O-H band. Such a variation associated with –OH stretching vibration proves the formation of hydrogen bonds, which is in good accordance in PVA/graphene nanocomposites [23] and PVA/BC nanocomposites [5]. The aforementioned results failed to show existing new bands in PVA films with the inclusion of both HNTs and Cloisite 30B clays. On the contrary, the inclusion of 3 and 5 wt% NBCs within PVA matrices in bionanocomposite films gives rise to a new band in relation to $-CH_2-$ asymmetric and symmetric stretchings [4]. Such a finding can arise from typical porous structures of NBCs, enabling to absorb molecular chains of hydrophilic polymers such as PVA with combined mechanical and chemical bondings. More consistently, the inclusion of HNTs and Cloisite 30B clays in PVA bionanocomposites shift the hydroxyl peaks of PVA to relatively high wavenumbers as opposed to the addition of NBCs [4]. Such results indicate that the numbers of hydrogen bonding generated in PVA/HNT bionanocomposites and PVA/Cloisite 30B clay bionanocomposites are higher when compared with those detected in PVA/NBC bionanocomposites owing to different chemical structures of nanofillers. Since NBCs do not possess –OH peaks, most hydrogen bonding formed in bionanocomposites can be associated with hydroxyl groups of PVA molecules. Whereas, the existing –OH peaks detected in HNTs and Cloisite 30B clays in PVA bionanocomposites are believed to further facilitate the formation of more hydrogen bonds within PVA matrices.

4.4 XRD Patterns

XRD analysis is a very useful material characterisation method to evaluate crystalline structures of polymers and composites, as well as to determine *d*-spacing values between clay interlayers. By monitoring the position and intensity of basal reflections from distributed silicate layers, nanocomposite structures (i.e. intercalated or exfoliated), as well as clay aggregated structures can be identified accordingly [24]. XRD patterns of HNTs and corresponding bionanocomposites are presented in Fig. 4.3(a). HNT patterns possess three major peaks of (001), (020)/(110) and (002) located at $2\theta = 11.9°$, 20° and 24.9°, corresponding to *d*-spacing values of 0.74, 0.44 and 0.37 nm, respectively. The relevant peak taking place at $2\theta = 24.9°$ is attributed to the presence of silica in form of cristobalite and quartz [25]. After the incorporation of HNTs into PVA matrices in nanocomposite systems, XRD characteristic peak at $2\theta = 11.9°$ appears to be very weak at the low HNT content of 3 wt%. Similar XRD peaks have been detected at $2\theta = 12.5°$ and 12.6° with a slight peak shift at relatively high HNT contents of 5 and 10 wt%, respectively. Overall,

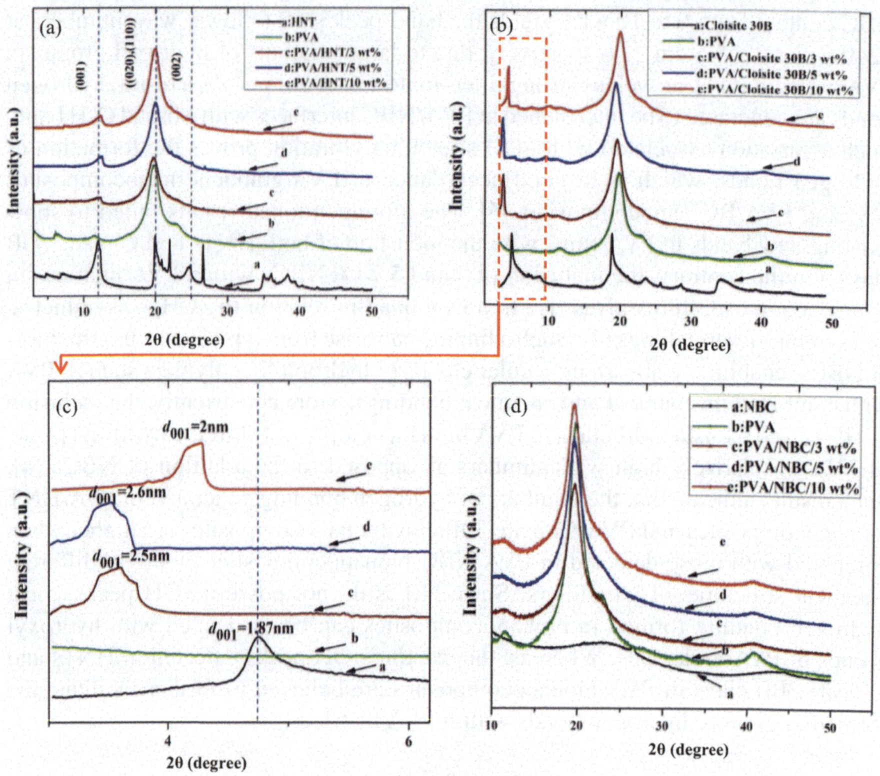

Fig. 4.3 XRD patterns for PVA bionanocomposites reinforced with **a** HNTs, **b**, **c** Cloisite 30B clays based on both wide and small diffraction angles, respectively, and **d** NBCs [12]

decreasing the HNT content significantly reduces the intensity of XRD peaks for all PVA/HNT bionanocomposites, which may result from uniform HNT dispersion at the low HNT content levels. Such a phenomenon suggests that homogeneous HNT dispersion in a more randomly-oriented manner may contribute to the disappearance or intensity reduction of XRD peaks at low HNT contents. On the other hand, the reappearance of XRD peaks at higher HNT contents is indicative of possible HNT agglomeration.

The XRD patterns of Cloisite 30B clays reveal the diffraction peak at $2\theta = 4.72°$ corresponding to the *d*-spacing value of 1.87 nm as shown in Fig. 4.3b, c. The (001) diffraction peak shifts to lower angles, as evidenced by *d*-spacing values of 2.5, 2.6 and 2 nm for PVA/Cloisite 30B clay bionanocomposites at the clay contents of 3, 5 and 10 wt%, respectively. It clearly arises from the diffusion of polymeric chains inside clay interlayers to create clay intercalation. Such a result is in good agreement with PVA/Na^+ MMT nanocomposite [16] and PLA/Cloisite 30B nanocomposites [26]. The XRD peak for PVA alone appears at $2\theta = 19.7°$, which is associated with total (101) crystalline atactic formation of PVA molecular chains [27] to slightly shift to lower diffraction angles when increasing the clay content in PVA bionanocomposites. The occurrence of PVA molecular chains at (101) crystalline plane suggests that PVA matrices evolve towards crystalline structures under more constraints. A similar behaviour has also been reported by Strawhecker and Manias [27] as well as Gaume et al. [28] in PVA/clay nanocomposites, which is ascribed to strong chemical interactions between nanofillers and polymer matrices. The aforementioned results indicate that Cloisite 30B clays are successfully intercalated and/or exfoliated into PVA molecular chains, and HNTs are homogenously dispersed at their low contents within continuous PVA matrices. This is attributed to close interactions of PVA matrices with clay nanoparticles due to strong hydrogen bonding taking place between carboxyl groups of PVA molecules and hydroxyl groups in the interlayer areas of Cloisite 30B clays or at the surface edges of HNTs [23].

In comparison, the XRD patterns of NBCs demonstrate two broad XRD peaks, as depicted in Fig. 4.3d. The broad peaks located at $2\theta = 22.9°$ are associated with sharp peaks of graphite assigned to (002) diffraction plane [29]. Besides, the second broad peak detected at $2\theta = 43.6°$ characterises 2D in-plane symmetry (101) along with graphene layers. Moreover, XRD patterns of PVA/NBC bionanocomposites only show the diffraction angles from PVA, as illustrated in Fig. 4.3(d). This phenomenon is consistent with previous finding in PVA/5 wt% graphene oxide (GO) nanocomposites [30] with clear disappearance of GO diffraction peaks in regular and periodic structures leading to individually exfoliated GOs within PVA matrices.

4.5 Topographic Surface Morphology and Surface Roughness

To assess the nanofiller dispersion within PVA matrices, 3D height mapping images of PVA and PVA bionanocomposites are exhibited in Fig. 4.4. As illustrated in Fig. 4.4b, one can see that HNT nanoparticles are separated from one another with better HNT dispersion in PVA bionanocomposites reinforced with 3 wt% HNTs, as opposed to typical clay agglomeration and clustering issues beyond 3 wt% HNTs shown in Fig. 4.4c, d. An excessive amount of HNTs results in decreasing intraparticle spacing along with higher intramolecular bonding of HNTs leading to particle agglomeration [31]. Besides, the average root-mean-square (R_q) value as an indicator of surface roughness of PVA bionanocomposites is reported to be 2.4 ± 0.13 nm at the HNT content of 3 wt% relative to 1.9 ± 0.17 nm for neat PVA, as shown in Fig. 4.4a. It is suggested that smooth surfaces of PVA/HNT bionanocomposites remain with the incorporation of HNTs at relatively low HNT contents owing to their uniform dispersion. Such a phenomenon is consistent with PVA/HNT composite hydrogels reported elsewhere [32]. At the low HNT content of 3 wt%, smooth surfaces of PVA/HNT bionanocomposites can also be ascribed to the combination of strong interactions and good compatibility between HNTs and PVA matrices. Nonetheless, increasing the HNT content up to 5 and 10 wt% induces much higher surface roughness (i.e. R_q = 4.54 ± 0.18 and 13.1 ± 0.23 nm, respectively) owing to the presence of prevalent HNT aggregates [33].

On the other hand, PVA/Cloisite 30B clay bionanocomposites and PVA/NBC bionanocomposites reveal different dispersibilities, as compared with PVA/HNT bionanocomposites. When nanofiller contents were below 10 wt%, spiky nanoparticles appeared to be separated from one another resulting in the homogeneous dispersion of Cloisite 30B clays and NBCs within PVA matrices, Fig. 4.4e, f, as well as Fig. 4.4h, i, respectively. In particular, as the nanofiller content increased from 3 to 5 wt%, R_q values increased moderately from 2.04 ± 0.12 to 2.84 ± 0.18 nm for PVA/Cloisite 30B clay bionanocomposites, as well as from 2.1 ± 0.11 to 2.5 ± 0.16 nm for PVA/NBC bionanocomposite, as opposed to 1.9 ± 0.17 nm for neat PVA. This finding suggested that smooth surfaces for PVA bionanocomposites were evident at the low nanofiller contents of Cloisite 30B clays and NBCs, which was in good agreement with previous study in PVA/nanocellulose composite films [34]. On the contrary, the inclusion of 10 wt% Cloisite 30B clays and NBCs in PVA bionanocomposites consistently gave rise to increasing R_q values up to 6.05 ± 0.23 and 4.1 ± 0.19 nm, respectively, which seemed also far higher than that of neat PVA at 1.9 ± 0.17 nm. Such results signified that the presence of aggregated Cloisite 30B clays and NBCs resulted in much higher surface roughness on PVA surfaces as expected. In comparison, the R_q value of 4.1 ± 0.19 nm for PVA/NBC bionanocomposites appeared to be relatively low, as compared with those of other PVA nanocomposites reinforced with carbon-based fillers such as PVA/reduced GO (rGO) nanocomposites with a R_q value of 4.6 ± 0.55 nm based on deposition layers [35].

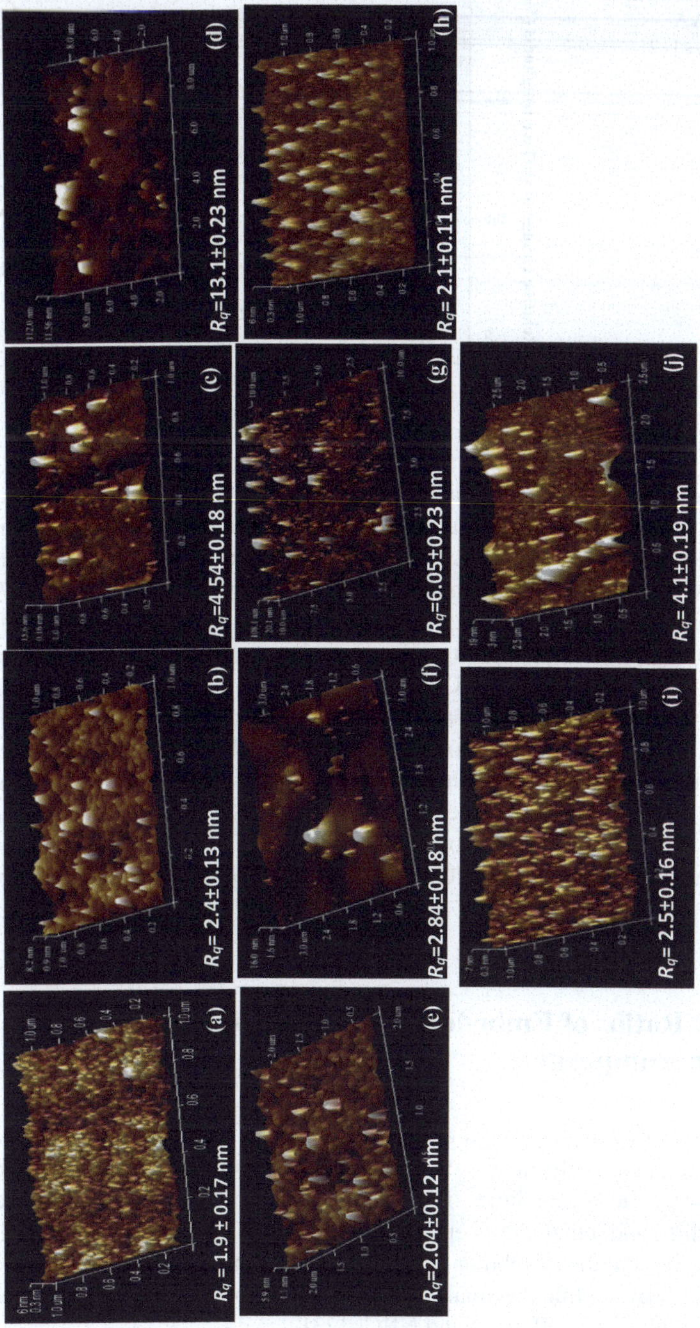

Fig. 4.4 3D height mapping images of **a** PVA and PVA bionanocomposites reinforced with **b** 3 wt% HNTs, **c** 5 wt% HNTs, **d** 10 wt% HNTs, **e** 3 wt% Cloisite 30B clays, **f** 5 wt% Cloisite 30B clays, **g** 10 wt% Cloisite 30B clays, **h** 3 wt% NBCs, **i** 5 wt% NBCs and **j** 10 wt% NBCs [12]

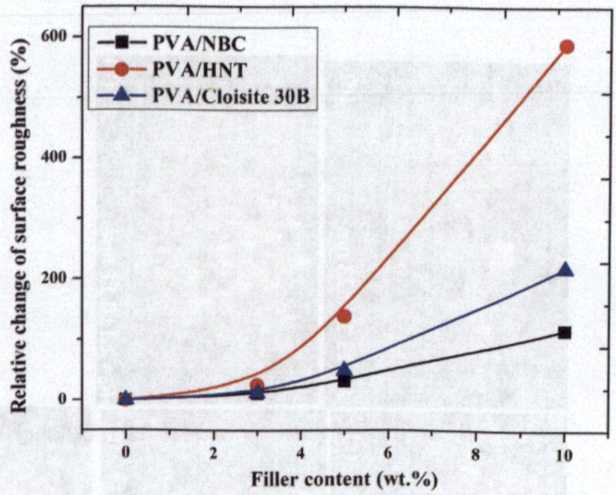

Fig. 4.5 Relative change of surface roughness in terms of filler content in PVA bionanocomposites [12]

Notwithstanding that the same manufacturing process condition and nanofiller contents have been utilised for preparing PVA bionanocomposite films, different nanoparticle types play an important role in the variation of the degree of roughness. Overall, with increasing the nanofiller content, the surface roughness of PVA bionanocomposites in this study was enhanced to different extent, as evidenced by increasing the maximum relative change of surface roughness up to 589.4, 218.4 and 115.8% with the inclusion of HNTs, Cloisite 30B clays and NBCs at the same filler content of 10 wt% shown in Fig. 4.5. Such a phenomenon suggested that NBCs might have better ability to be dispersed more uniformly in PVA matrices, as opposed to HNTs and Cloisite 30B clays due to their least increasing level in surface roughness especially when beyond 5 wt% in filler content.

While the effect of different nanoparticle shapes and sizes on surface roughness of PVA bionanocomposites became less pronounced at low filler contents below 3 wt%.

4.6 Aspect Ratios of Embedded Nanofillers in PVA Bionanocomposites

The aspect ratio of nanofillers is regarded as one of key factors in reinforcement efficiency and mechanical performance of nanocomposites, which is generally defined as the ratio between the largest dimensions over the smallest dimension of nanofillers. According to this fundamental concept, the largest dimension of nanofillers can be represented by the lengths of tubular HNTs and platelet-like Cloisite 30B clays or the diameters of NBCs while the smallest dimension denotes the diameter of HNTs or thicknesses of Cloisite 30B clays and NBCs [23].

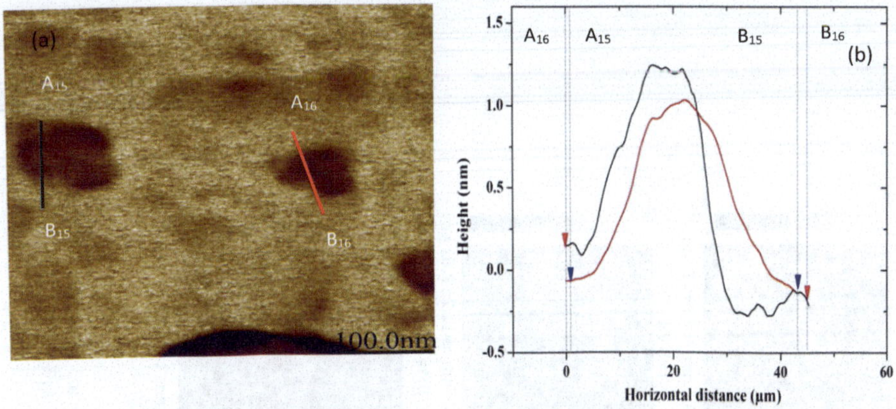

Fig. 4.6 AFM characterisation of PVA/3wt% Cloisite 30B bionanocomposites. **a** Height mapping image and **b** corresponding profiles for cut sections A_{15}-B_{15} and A_{16}-B_{16} [12]

For instance, it is well known that when nanoclays are uniformly dispersed within polymer matrices, the formation of their exfoliated and intercalated structures leads to the improvement of mechanical performance of nanocomposites to different extent, which is opposed to agglomerated nanoclays resulting in the deterioration of their mechanical properties [1]. To investigate the degree of clay exfoliated structures in details, the height profiles of clay platelets relative to that of PVA matrices have been determined, as shown in Fig. 4.6. The thickness of 3 wt% Cloisite 30B clays within PVA matrices in bionanocomposites appears to be in a range of 0.85–1.43 nm, suggesting typical exfoliated clay structures in filler dispersion. MMT clays are well known to be exfoliated when their thickness is similar to that of individual clay platelet (i.e. ~1 nm) [13]. Gaume et al. [28] and co-workers [36, 37] also detected intercalated and exfoliated structures of MMT clays in the thickness range of 1.3–5 nm.

Surface roughness mentioned earlier can be associated with nanofiller shape and size since HNTs and Cloisite 30B clays may possess relatively high aspect ratios when compared with that of NBCs with existing 'nanofiller waviness' issue. High-aspect-ratio HNTs and Cloisite 30B clays inevitably undergo considerably wavy nanofiller formation, thus undermining their homogeneous dispersion within polymer matrices [23]. In addition, nanofiller dispersion techniques used such as ultrasonication may also potentially damage nanofiller structures if high power intensity or longer sonication time applies [38]. Hence, specific dimensions of nanofillers required to calculate their actual aspect ratios were determined in this study for embedded NBCs, HNTs and Cloisite 30B clays in PVA bionanocomposites, as illustrated in Fig. 4.7. The detailed frequency distributions of nanofiller dimensions are presented in Figs. 4.8, 4.9 and 4.10. It is clearly revealed that aspect ratios of nanofillers increased from 5.91 to 10.60 for HNTs in Fig. 4.7a–c, as well as 5.75–8.17 for NBCs in Fig. 4.7g–i with increasing the nanofiller content from 3 to 10 wt%. In contrast, the aspect ratios of Cloisite 30B clays decreased from 22.70, 12.38 to 13.46 when increasing the clay contents from 3, 5 to 10 wt% accordingly despite their overall highest aspect ratios

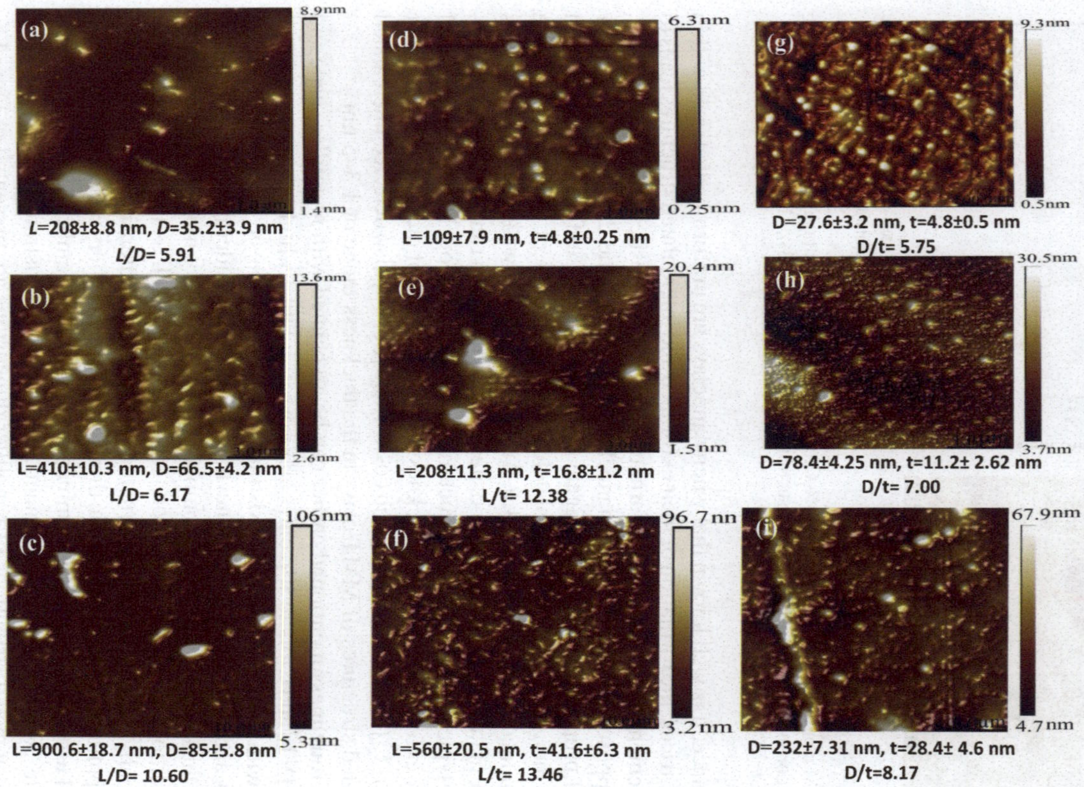

Fig. 4.7 AFM topographic images of PVA bionanocomposites associated with aspect ratios of embedded fillers: **a** 3 wt% HNTs, **b** 5 wt% HNTs, **c** 10 wt% HNTs, **d** 3 wt% Cloisite 30B clays, **e** 5 wt% Cloisite 30B clays, **f** 10 wt% Cloisite 30B clays, **g** 3 wt% NBCs, **h** 5 wt% NBCs and **i** 10 wt% NBCs [12]

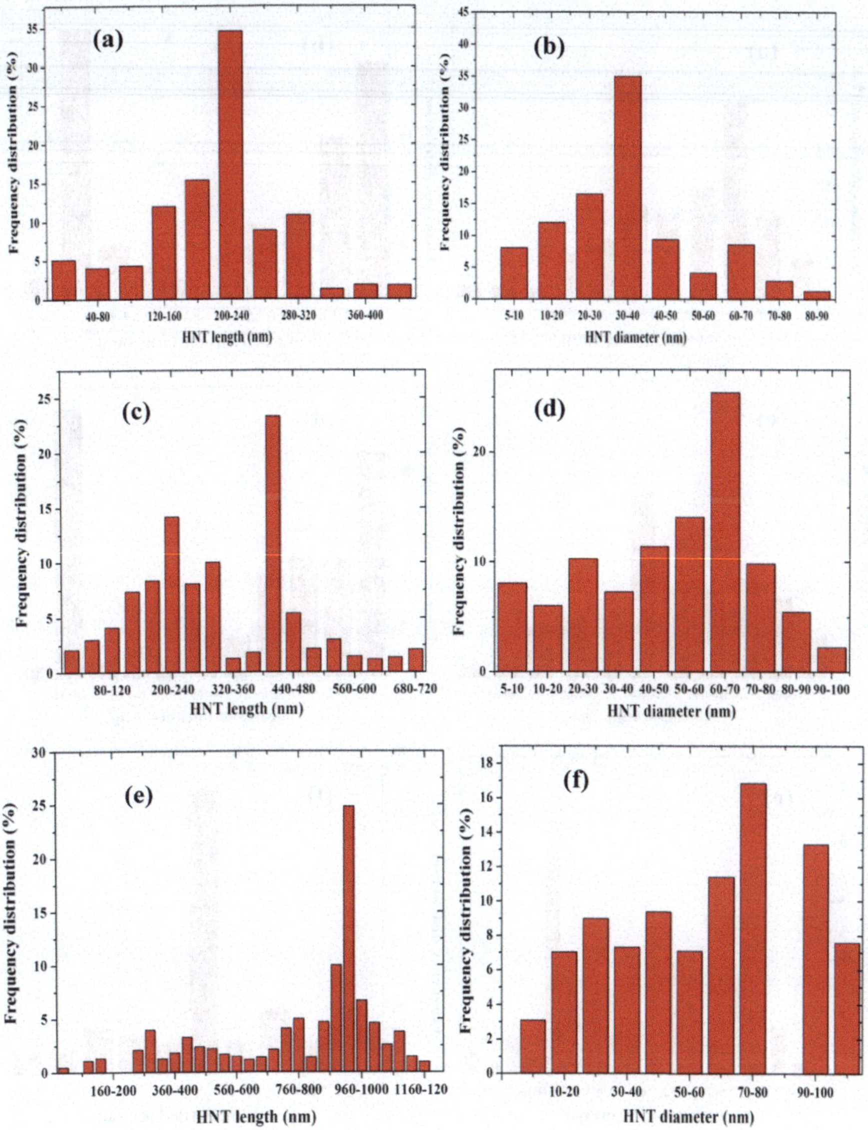

Fig. 4.8 Frequency distributions of HNT length and diameter in PVA/HNT-based bionanocomposites at different HNT contents: **a**, **b** 3 wt%, **c**, **d** 5 wt% and **e**, **f** 10 wt% [12]

among all nanofillers, Fig. 4.7d–f. Such findings implied that the majority of Cloisite 30B clays tended to form exfoliated or intercalated clay structures with relatively high aspect ratios. However, the decrease in aspect ratios of Cloisite 30B clays could be associated with more severe clay aggregation. It is very convincing that aspect ratios of nanofillers can be greatly influenced by nanofiller shapes, and apparently

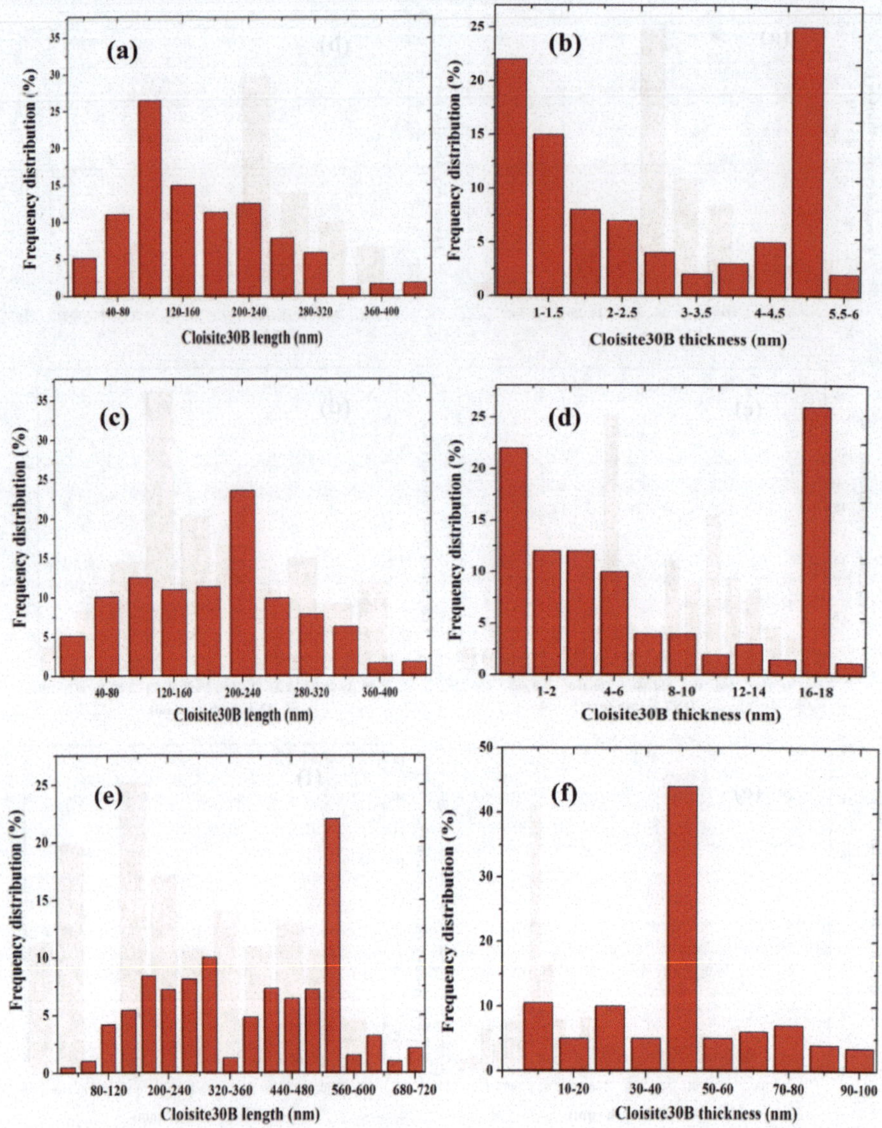

Fig. 4.9 Frequency distributions of Cloisite 30B clay length and thickness in PVA/Cloisite 30B clay bionanocomposites at different clay contents: **a**, **b** 3 wt%, **c**, **d** 5 wt% and **e**, **f** 10 wt% [12]

3D NBC nanoparticles had relatively low aspect ratios when compared with 1D platelet-like Cloisite 30B clays [23].

Overall, aspect ratio may play a significant role in mechanical performance of nanocomposites when nanofiller shape and structure are only considered within polymer matrices. However, for different types of nanofillers, several other factors

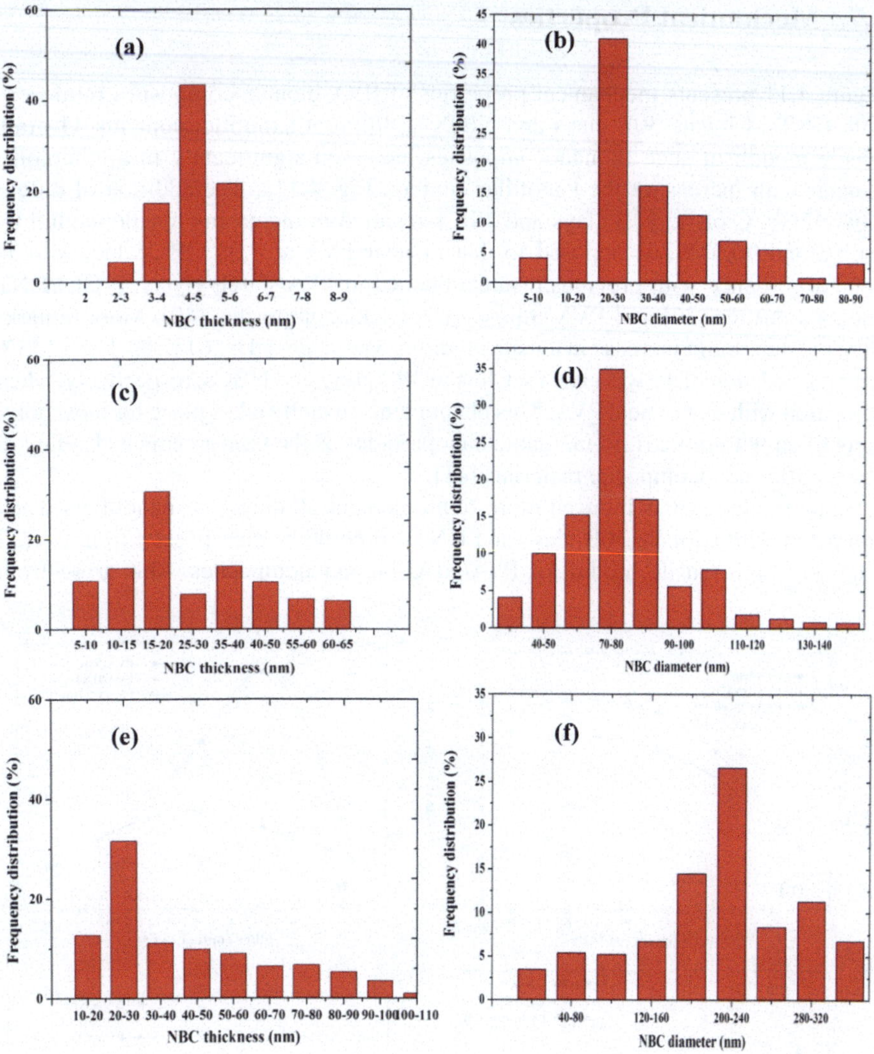

Fig. 4.10 Frequency distributions of NBC length and diameter in PVA/NBC bionanocomposite at different NBC contents: **a**, **b** 3 wt%, **c**, **d** 5 wt% and **e**, **f** 10 wt% [12]

like the number of particles per unit volume, interphase modulus, interphase volume and surface area, as well as the ratio of interphase volume per nanoparticle volume should also be taken into account for overall material properties.

4.7 Mechanical Properties

Figure 4.11 presents mechanical properties of PVA bionanocomposites reinforced with HNTs, Cloisite 30B clays and NBCs at different nanofiller contents. Overall, tensile moduli of such bionanocomposites increased significantly in a monotonic manner with increasing the nanofiller content, Fig. 4.11a. The addition of only 3 wt% HNTs, Cloisite 30B clays and NBCs resulted in increasing tensile moduli by 40, 52 and 70.67%, as opposed to that of near PVA at 2.08 GPa, which was in good accordance with previous results obtained in PVA/starch/glycerol (GL)/HNT nanocomposites [39] and PVA/chitosan/HNT nanocomposites [40]. More remarkably, the maximum increase in tensile modulus was achieved by 61.5, 84.1 and 123% with the addition of 10 wt% HNTs, Cloisite 30B clays and NBCs, respectively, when compared with that of neat PVA. This phenomenon usually takes place for most polymers filled with more rigid inorganic nanoparticles as the reinforcements leading to much stiffer nanocomposite materials [41].

In particular, NBCs induced more reinforcement efficiency as nanofillers when compared with Cloisite 30B clays and HNTs. It could be clearly seen from overall relatively high tensile moduli of PVA/NBC bionanocomposites. This arose from

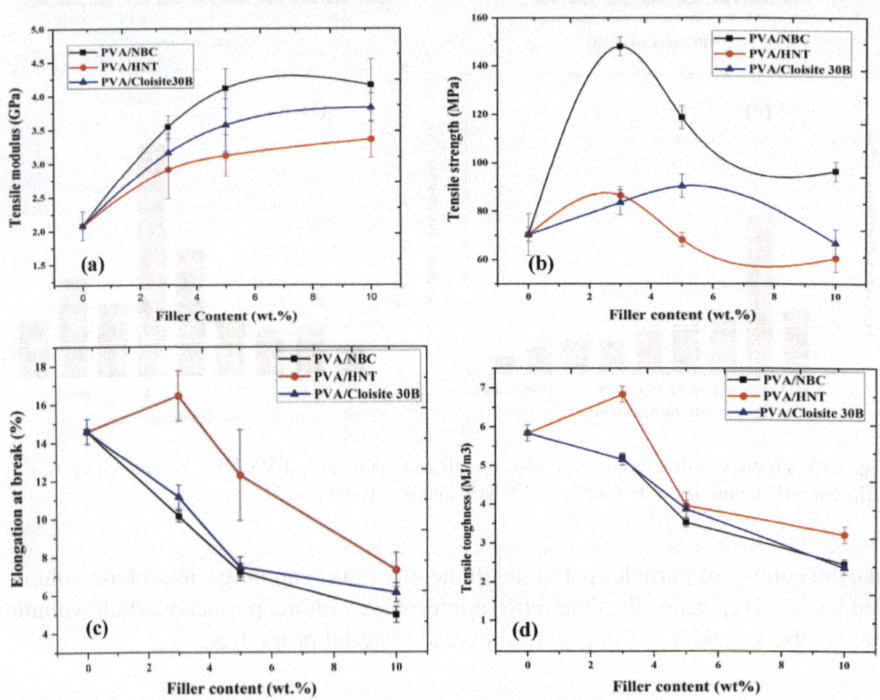

Fig. 4.11 Mechanical properties of PVA bionanocomposites at different filler contents: **a** tensile modulus, **b** tensile strength, **c** elongation at break and **d** tensile toughness [12]

much closer interactions between PVA matrices and NBCs via the formation of mechanical and hydrogen bonds owing to highly porous structures of NBCs. On the other hand, a different trend for tensile strengths of PVA/NBC bionanocomposites was revealed from those of PVA/Cloisite 30B clay bionanocomposites and PVA/HNT bionanocomposites, as shown in Fig. 4.11b. The tensile strength of PVA/HNT bionanocomposites was improved by 23% with the addition of 3 wt% HNTs relative to that of neat PVA at 70.32 MPa. Nonetheless, a drastic strength-decreasing tendency occurred with the strength reductions by 3.2 and 13.9% when embedded with 5 and 10 wt% HNTs, respectively. Such results indicated that the enhancement of tensile strengths for PVA/HNT bionanocomposites depended on effective stress transfer from PVA matrices to HNTs, mostly resulting from homogeneous HNT dispersion within PVA matrices. On the contrary, increasing the HNT content inevitably gave rise to noticeable particle aggregation with more stress concentration sites around HNT agglomerates as a result of potential crack initiation to deteriorate mechanical performance of such bionanocomposites. With respect to PVA/Cloisite 30B clay bionanocomposites, their tensile strengths were increased by 18.8 and 28.4% with the incorporation of 3 and 5 wt% Cloisite 30B clays, respectively, which was ascribed to more uniform clay dispersion, as well as the formation of stronger matrix–filler network structures, resulting from increasing hydrogen bonding between these constituents due to larger clay surface areas [42, 43]. When Cloisite 30B clay content increased up to 10 wt%, the tensile strength of PVA/Cloisite 30B clay bionanocomposites was decreased by 5.16% as opposed to that of near PVA. This finding suggested that the aggregation of nanofillers at high clay content levels could diminish tensile strength. On the contrary, the tensile strength of PVA/NBC bionanocomposites possessed the initial improvement up to 147.94 MPa (by a maximum level of 110.4%) when the NBC content increased from 0 to 3 wt%. Beyond 3 wt% NBCs, tensile strengths of such bionanocomposites tended to decline until they reached the lowest strength levels of 96.34 MPa at the NBC content of 10 wt% despite being still better than that of neat PVA. Overall, both tensile moduli and tensile strengths of PVA/NBC bionanocomposites were consistently superior to those of PVA/HNT bionanocomposites and PVA/Cloisite 30B clay bionanocomposites, thus validating the most effective reinforcement efficiency of NBCs among all three different nanofillers.

The elongation at break and tensile toughness of PVA/Cloisite 30B clay bionanocomposites and PVA/NBC bionanocomposites continuously decreased especially beyond the nanofiller content of 3 wt%, Fig. 4.11c, d. The maximum decreasing levels by approximately 59.5 and 58% in elongation at break were detected for PVA bionanocomposites reinforced with 10 wt% Cloisite 30B clays and NBCs, respectively. This finding can be associated with the stiffening effect from filler reinforcements of NBCs and Cloisite 30B clays to restrict the movement of PVA molecular chains, thus resulting in the overall reduction in the flexibility of bionanocomposite films [44]. As for PVA/HNT bionanocomposites, elongation at break and tensile toughness were increased by 12.7 and 16.9% with the incorporation of 3 wt% HNTs, beyond which both of them remarkably diminished until the maximum reductions of 50 and 45.3% took place at the HNT content of 10 wt%, respectively, as opposed to those of PVA. The former finding can be explained by good particle–matrix interactions with more uniform particle dispersion at low HNT contents. Whereas, the latter result can be associated with typical particle agglomeration at high HNT contents up to 10 wt% with the disappearance of 'nanoeffect' of HNTs since most HNT aggregates become less favourable microfillers with poor particle dispersion. As such, those HNT aggregates could act as typical defects with high stress concentration prone to crack initiation towards mechanical failure, thus leading to poor material toughness [31].

In this study, the incorporation of three different nanofillers (i.e. HNTs, Cloisite 30B clays and NBCs) successfully enhanced mechanical properties of PVA bionanocomposite films. According to our results, the highest increasing level among PVA/HNT nanocomposites and PVA/NBC nanocomposites could be achieved at the filler content of 3 wt%, as opposed to the optimum content of 5 wt% for PVA/Cloisite 30B clay bionanocomposites. Nonetheless, the increasing rate achieved in PVA bionanocomposites using three nanofillers appeared to be quite different, which could be associated with a variety of nanofiller features in terms of their structures and geometries, as well as the degree of compatibility between nanofillers and polymer matrices. With respect to nanofiller shape, it is well known that NBCs are regarded as 3D nanofillers as opposed to 2D nanofillers for HNTs and 1D nanofillers for Cloisite 30B clays. Different nanofiller shapes thereby influence the overall interfacial areas between fillers and polymer matrices, which plays a key role in the improvement of tensile strengths of nanocomposites with different filler–matrix interactions. The second aspect is related to the structures, particularly for the location of hydroxyl groups with respect to nanofiller structures and the amounts of hydroxyl groups within nanofillers. In case of NBCs, hydroxyl groups are located inside their pores, which tend to more closely interact with PVA in a 3D view. As for HNTs, the majority of hydroxyl groups is constrained in inner tubes between layers, which make the matrix-HNT interaction limited to inner tubes of HNTs only. Moreover, in case of Cloisite 30B clays, hydroxyl groups are located between layered structures, which means that the interactions between polymer matrices and platelet-like clays are limited to small constrained interlayer areas. Highest tensile moduli and tensile strength of bionanocomposite films have been achieved with the incorporation of NBCs relative to those of PVA. Several reasons could explain the above-mentioned results in relation to mechanical properties of bionanocomposites. First, 3D nanofiller shape of NBCs can be generated at low nanofiller contents and in small particle sizes with relatively large interfacial areas as compared with 2D HNTs and 1D Cloisite 30B clays. Liu and Brinson [2] investigated the effect of nanofiller geometry on the reinforcing efficiency of nanocomposites, which showed that at a low nanofiller content with the random nanofiller orientation, the transverse modulus of nanoparticle-based nanocomposites significantly exceeded those of nanotube-based nanocomposites, as well as nanoplatelet-based nanocomposites. Schadler et al. [45] reported that in case of a nanocomposite system with the incorporation of nanoparticles and nanotubes having a nanofiller diameter of 10 nm at the volume fraction of 10 vol%, the volume fraction of interfacial polymers was found to be about 30% in case of nanoparticle-based nanocomposites, as opposed to only 10% for nanotube-based nanocomposites. The second reason in relation to high mechanical performance of PVA/NBC bionanocomposites could be ascribed to chemical structures of nanofillers in terms of the amounts and locations of hydroxyl groups in order to control the nanofiller dispersion within bionanocomposites, thus reflecting upon the bonding between polymer matrices and nanofillers. Pakzad et al. [46] reported that the number and nature of hydrogen bonds had substantial effect on mechanical properties of nanocomposites. In case of PVA/3 wt% NBC bionanocomposites, NBCs have highly porous structures with a large amount of hydroxyl groups located inside these pores when NBCs are uniformly dispersed. As confirmed by FTIR and XRD results, polymeric chains enter these pores and form both hydrogen and mechanical bondings. Such two bonding types can be particularly recognised for NBCs, as compared to Cloisite 30B clays and HNTs, thus leading to significantly enhanced mechanical properties of nanocomposites [4, 5]. In case of PVA/5 wt% Cloisite 30B clay bionanocomposites in comparison to PVA/5 wt% HNT bionanocomposites, the strong adhesion of clays in polymer matrices associated with uniform clay dispersion gave rise to the strong interfacial bonding between nanoclays and polymer matrices, which thus could predominantly contribute to the improvement of mechanical properties of nanocomposites.

The dispersion state of nanofillers can also influence mechanical properties of PVA bionanocomposites. As mentioned earlier, NBCs have better dispersion state than Cloisite 30B clays and HNTs. The incorporation of NBCs within PVA matrices produces smoother bionanocomposite films with higher tensile strength when compared with those of PVA/HNT bionanocomposites and PVA/Cloisite 30B clay bionanocomposites. The better dispersion state of NBCs improves their interaction with PVA matrices, thus resulting in higher tensile strengths of PVA/NBC bionanocomposites. On the contrary, increasing the nanofiller content appears to induce higher surface roughness, as well as lower tensile strengths of nanocomposites, which indicates nanofiller agglomeration apparently has detrimental effect on the improvement of tensile strength. This is particularly the case for PVA/HNT bionanocomposites due to poor HNT dispersion state. Conversely, PVA/3 wt% HNT bionanocomposites yield much higher elongation at break and fracture toughness, as opposed to those of neat PVA, which are different from PVA/3 wt% NBC bionanocomposites and PVA/5 wt% Cloisite 30B clay bionanocomposites with corresponding lower values. Such results can be clearly explained by two major reasons. The first reason is ascribed to the number of nanoparticles that depend on the volume and volume fraction of nanoparticles in nanocomposites. At the same volume fraction, the number of near spherical NBCs is significantly larger than those of tubular HNTs or platelet-like Cloisite 30B clays. As such, this finding resulted in increasing the number of available reinforcements for improving matrix rigidity, and then decreasing fracture toughness [38, 47]. The second reason is related to the mechanism of fracture toughness including pre-crack effect for the fracture of nanocomposites. In general, crack deflection and crack pinning are most well-known mechanisms resulting in an increase in fracture energy [48], and consequently an increase in fracture toughness of nanocomposites. In both mechanisms mentioned earlier, crack growth path can increase as long as those cracks reach nanofiller regions and the reinforcement shape highly affects the amount of crack deviation from their initial path. Since HNTs have larger lateral dimensions in comparison with near spherical NBCs, the cracks tend to pass over longer distances in PVA bionanocomposites reinforced with HNTs. Moreover, crack bridging is a well-known fracture mechanism in nanocomposites reinforced with high-aspect-ratio nanoparticles [25]. An ideal situation in this mechanism occurs when nanotube fillers are still embedded in matrices while aligned in a perpendicular direction to crack faces. Consequently, their higher aspect ratios yielded the improvement of fraction toughness when compared with those of near spherical nanoparticles.

4.8 Fracture Morphology

Figure 4.12 shows typical SEM micrographs of cross-sectional fracture surfaces for PVA, PVA/HNT bionanocomposites, PVA/Cloisite 30B clay bionanocomposites and PVA/NBC bionanocomposites. It can be clearly seen in Fig. 4.12b, e that PVA bionanocomposites reinforced with 3 wt% of HNTs and Cloisite 30B clays reveal much rougher fractured surfaces when compared with those of neat PVA films, as illustrated in Fig. 4.12a. Moreover, 3 wt% HNTs or Cloisite 30B clays are distributed uniformly within PVA matrices. The good dispersion of both nanoparticles and strong interaction between clay particles and polymer matrices clearly contribute to the reinforcing effect, as a result of an increase in both tensile strength and elastic modulus. Nevertheless, in PVA/HNT bionanocomposite and PVA/Cloisite 30B clay bionanocomposite systems, uniform multi-layered structures have not been achieved similar to those detected in PVA/3 wt% NBC bionanocomposites illustrated in Fig. 4.12h. Such results are indicative of high NBC dispensability relative to those of HNTs and Cloisite 30B clays along with the highest mechanical performance. Meanwhile, at the HNT content of 5 wt%, particle–particle interactions are more favourable than particle–matrix counterparts, as evidenced by more filler agglomeration in presence of debonding and microvoid effects depicted in Fig. 4.12c. Such defects in nanocomposite systems give rise to the decrease in tensile strengths of PVA/HNT bionanocomposites. However, as for PVA/5 wt% Cloisite 30B clay bionanocomposites, the clay dispersion appears to be still relatively uniform with presence of small particle agglomeration shown in Fig. 4.12f. With increasing the nanofiller contents of HNTs and Cloisite 30B clays from 5 to 10 wt%, the fracture surfaces of bionanocomposites films are altered from ductile characteristic to more brittle behaviour, as illustrated in Fig. 4.12d, g, respectively. Similar phenomena were also found in PVA/NBC bionanocomposites in Fig. 4.12j. It is well known that decreasing surface roughness can make the failure mode of PVA bionanocomposite films vary from ductile to brittle fracture [49], which is consistent with the reduced mechanical properties of bionanocomposite films in this study.

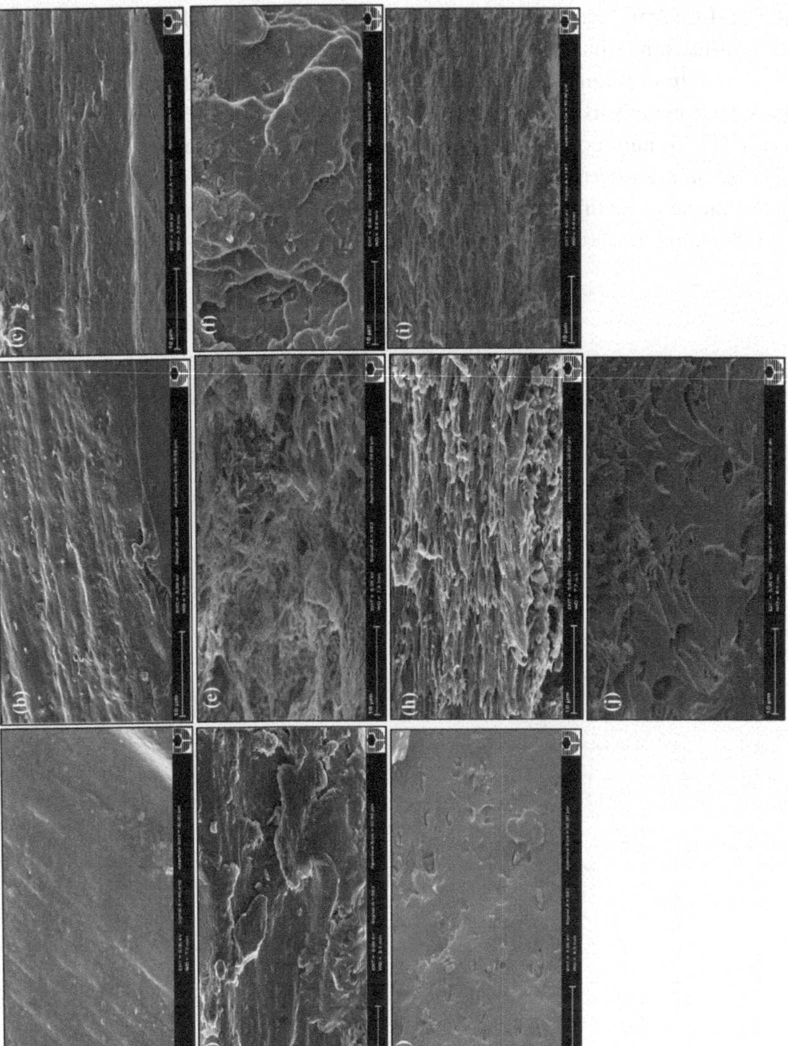

Fig. 4.12 SEM micrographs of tensile fracture surfaces: **a** PVA, **b** PVA/3 wt% HNT bionanocomposites, **c** PVA/5 wt% HNT bionanocomposites, **d** PVA/10 wt% HNT bionanocomposites, **e** PVA/3 wt% Cloisite 30B clay bionanocomposites, **f** PVA/5 wt% Cloisite 30B clay bionanocomposites, **g** PVA/10 wt% Cloisite 30B clay bionanocomposites, **h** PVA/3 wt% NBC bionanocomposites, **i** PVA/5 wt% NBC bionanocomposites and **j** PVA/10 wt% NBC bionanocomposites [12]

4.9 Thermal Properties

PVA is a water-soluble semicrystalline polymer, in which high physical interchain and intrachain interactions exist because of typical hydrogen bonding between hydroxyl groups. The inclusion of nanoclays with hydroxyl groups could alter the intramolecular and intermolecular interactions of PVA molecular chains. This may affect both crystallisation behaviour and physical structures of PVA. Similar observations can be found in previous studies dealing with PVA/HNT bionanocomposites [41, 50] and PVA/Cloisite 30B clay bionanocomposites [23]. Figure 4.13 shows the results of PVA/HNT bionanocomposites, PVA/Cloisite 30B clay bionanocomposites and PVA/NBC bionanocomposites determined by differential scanning calorimetry (DSC). The summarised data of these thermal characteristics are reported in Table 4.1. For PVA/HNT bionanocomposites, it is clearly seen that the T_g of PVA becomes

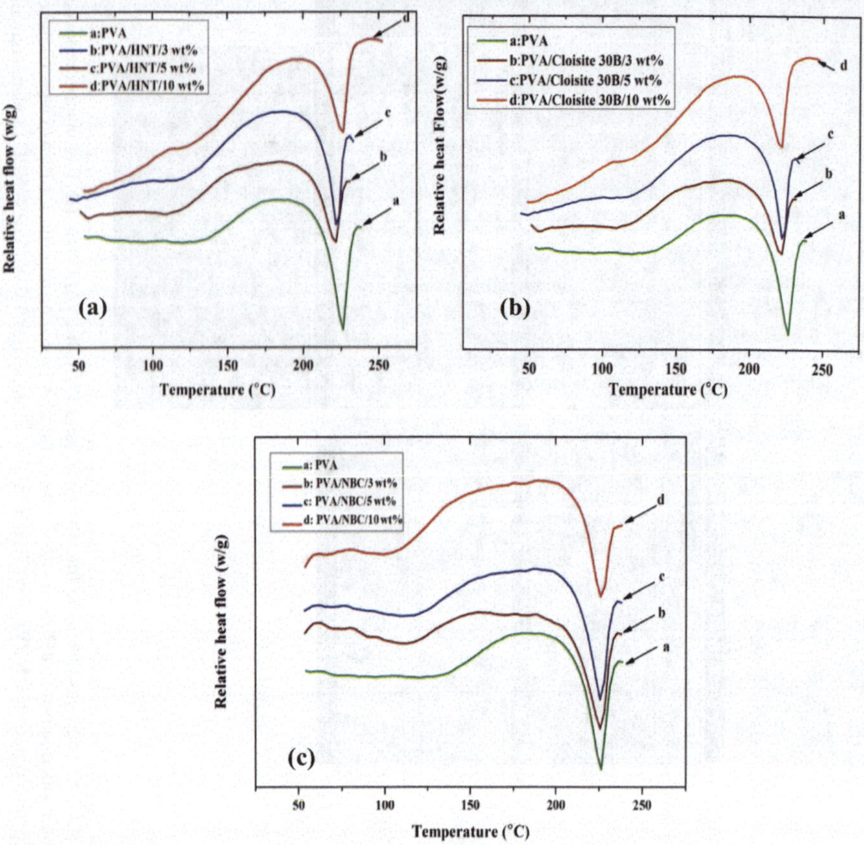

Fig. 4.13 DSC thermograms of PVA bionanocomposites reinforced with **a** HNTs, **b** Cloisite 30B clays and **c** NBCs [12]

Table 4.1 Thermal properties of different PVA bionanocomposite films [12]

Sample	T_g (°C)	ΔH_m (J/g)	χ_c (%)	T_m (°C)	$T_{5\%}$ (°C)	$T_{80\%}$ (°C)	T_d (°C)	T'_d (°C)
PVA	65.19	50.8	36.65	222.91	200.15	363.5	274.23	
PVA/HNT/3 wt%	65.13	50.5	37.6	224.74	265.29	387	283.2	
PVA/HNT/5 wt%	65.86	50.2	38.2	225.13	268.06	422.87	286.47	
PVA/HNT/10 wt%	64.91	44.15	35.4	226.67	270.21	428.03	287.13	
PVA/Cloisite30B/3 wt%	67.5	49.6	36.9	220.44	253.3	387.9	282.21	
PVA/Cloisite30B/5 wt%	70.2	49.09	37.2	221.62	261.1	407.12	287.41	
PVA/Cloisite30B/10 wt%	71.8	45.01	36.1	221.06	265.3	435.3	290.13	
PVA/NBC/3 wt%	70.53	50.01	37.21	222.12	256.3	390.67	278.48	
PVA/NBC/5 wt%	73.46	49.77	38.83	221.57	262.96	440.28	283.4	303.77
PVA/NBC/10 wt%	75.06	48.16	40	222.63	270.73	464.03	294.61	324.12

Note T_g, ΔH_m, χ_c and T_m are glass transition temperature, melting enthalpy, degree of crystallinity and melting temperature determined in DSC analysis, respectively. $T_{5\%}$, $T_{80\%}$, T_d and T'_d are the temperatures at the mass losses of 5 and 80%, as well as the first and second maximum decomposition temperatures in TGA accordingly unchanged with the addition of HNTs in bionanocomposite films, indicating that HNTs do not play an important role in inhibiting the chain mobility of PVA molecules. Qiu and Netravali [41] also reported a similar result in T_g with the incorporation of HNTs into PVA. Such a finding might be related to the reduction in the entanglements and interactions of PLA polymeric chains with HNT inclusions. The relatively unchanged T_g in PVA/HNT bionanocomposites, as compared to that of neat PVA, may arise from different nanofiller geometries. The diameters of HNTs are at a nanoscaled level as opposed to microsized tubular lengths that considerably exceed typical gyration radii of polymeric chains [4]. As a result, HNTs cannot be completely wrapped by PVA molecular chains leading to many voids surrounding HNT particles. On the contrary, high T_g values for all PVA/Cloisite 30B clay bionanocomposites are evidently shown as opposed to that of neat PVA. With the incorporation of 3, 5 and 10 wt% Cloisite 30B clays, T_g values of such PVA bionanocomposites are moderately enhanced up to 67.5, 70.2 and 71.8 °C, respectively, when compared with that of neat PVA at 65.19 °C. This phenomenon can be attributed to the confinement of polymeric chains by intercalated clay structures to prevent their segmental motions [1], which has also been reported in PVA/MMT nanocomposites [51, 52], PVA/bentonite nanocomposites [53], as well as PVA/starch/MMT nanocomposites [54]. In case of PVA/NBC bionanocomposites, the T_g increased monotonically up to 75.06 °C with increasing the NBC content from 0 to 10 wt% accordingly. The incorporation of rigid NBC particles can restrict the chain mobility of PVA matrices so that higher T_g values are required for the phase change of nanocomposites from a glassy state to a rubbery state. This finding is well known for many types of nanofillers such as nanoclays, GOs, CNTs and HNTs [1].

The degree of crystallinity (χ_c) of PVA slightly increased from 36.65% for neat PVA to 38.2, 37.2 and 40% for corresponding bionanocomposites with the incorporation of 5 wt% of HNTs and Cloisite 30B clays and 10 wt% of NBCs, respectively. This suggests that such nanofillers have minor effect on crystalline phases of PVA matrices in bionanocomposites. On the other hand, the melting temperature T_m of PVA bionanocomposites virtually has no change with the addition of Cloisite 30B clays and NBCs, as evidenced by the given T_m ranges of 220.44–221.62 °C and 221–225 °C, respectively, when compared with that of neat PVA at 222.91 °C. However, PVA/HNT bionanocomposites possess a moderate increase in T_m up to 226.67 °C with the inclusion of 10 wt% HNTs. A similar phenomenon was also noticed in PHBV/HNT nanocomposites [23] with their T_m values being increased from 169 to 173 °C when incorporated with 5 wt% HNTs. It is believed that based on XRD results in which HNTs were embedded within PHBV matrices, thicker and more oriented HNT/PHBV structures could be formed leading to higher melting temperatures.

Thermal decomposition behaviour of PVA/HNT bionanocomposites, PVA/Cloisite 30B clay bionanocomposites and PVA/NBC bionanocomposites was evaluated using a thermogravimetric analysis (TGA) with results being presented in Figs. 4.14 and 4.15, as well as Table 4.1. Relevant results for both systems reveal the existence of three major degradation steps according to previous studies [16]. Initially, the first degradation takes place at 107 °C owing to the breakage of hydrogen bonds, impurities and monomers of vinyl alcohol. Then the

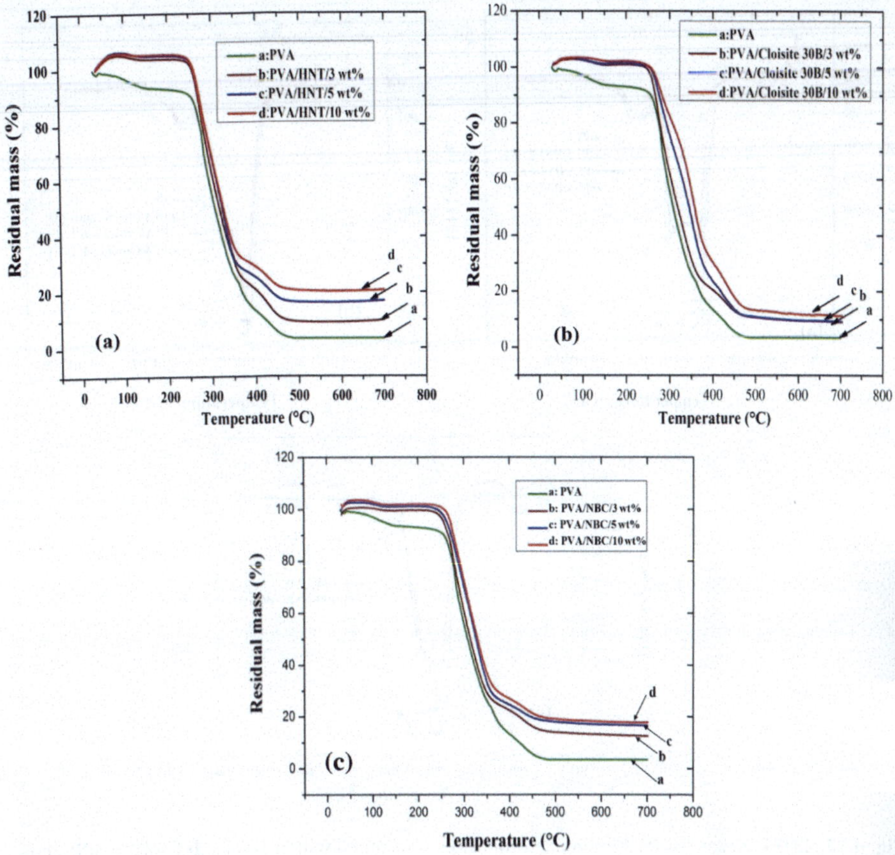

Fig. 4.14 TGA curves for PVA bionanocomposites reinforced with **a** HNTs, **b** Cloisite 30B clays and **c** NBCs [12]

second degradation taking place at 274 °C involves a dehydration reaction on PVA molecular chains, degradation of main backbones and decomposition of organic clays. This process is accompanied by a drastic mass change caused by the removal of organic compounds like CO_2 and long molecular chains of alkyl derivatives. Finally, the third degradation step appears at a temperature level below 429 °C with more complexity including the further degradation of polyene residues to yield the carbon and hydrocarbon. The incorporation of HNTs, Cloisite 30B clays and NBCs can increase the thermal stability of PVA by reducing the weight loss and increasing the decomposition temperatures, as presented in Figs. 4.14 and 4.15. For PVA/HNT bionanocomposites, the decomposition temperature at 5% weight loss increased from 200.2 °C for PVA to 265.3, 268.1 and 270.2 °C for PVA bionanocomposites reinforced with 3, 5 and 10 wt% HNTs, respectively. Such a finding suggests that HNTs act as an effective barrier material to heat or mass transfer. Moreover, intrinsic hollow tubular structures of HNTs can produce the traps for volatile particles, thus

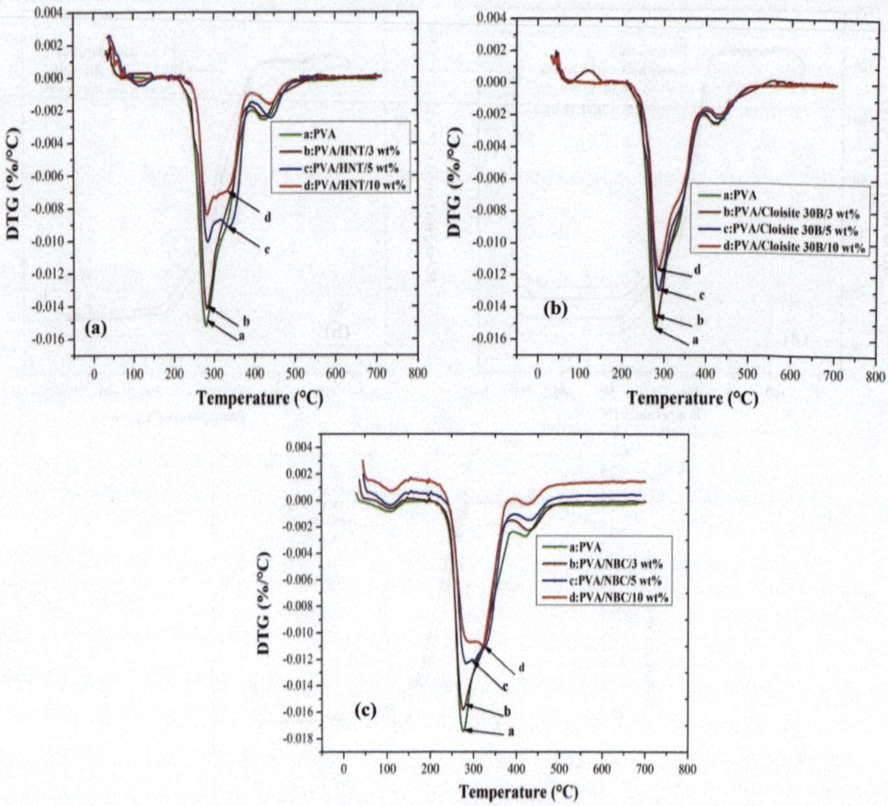

Fig. 4.15 DTG curves for PVA bionanocomposites reinforced with **a** HNTs, **b** Cloisite 30B clays and **c** NBCs [12]

improving thermal stability by delaying the mass transfer during a decomposition process. Moreover, as clearly seen from derivative thermogravimetric (DTG) curves in Fig. 4.15a, the T_d of PVA shifts to a higher temperature level, which means that the dehydration process is hindered, resulting from strong interactions between PVA matrices and HNTs, as well as HNTs also work as good barrier materials to increase the thermal resistance of PVA bionanocomposites. Furthermore, the second DTG peaks of PVA/HNT bionanocomposites reinforced with 5 and 10 wt% HNTs are much wider in the presence of main and side peaks, as compared to single DTG peak for PVA at the same step. This signifies that a single peak for PVA can be attributed to an eliminated reaction while side and main peaks of PVA/HNT bionanocomposites correspond to the eliminated reaction as well as the overlap of continual elimination and chain-scission reaction with the requirement of more energy to accrue at high temperatures [55].

On the other hand, in case of PVA/Cloisite 30B clay bionanocomposites, with increasing the clay content, the thermal stability of bionanocomposites improves, as compared to that of PVA, which is evidently noted from consistently high $T_{5\%}$, $T_{80\%}$ and T_d values shown in Fig. 4.14b. For instance, $T_{80\%}$ of PVA increases from 363.5 to 407 °C with the inclusion of 5 wt% Cloisite 30B clays. Such a result is in good agreement with previous studies of PVA/MMT nanocomposites [16, 56]. Moreover, the shift in the decomposition temperatures T_d for PVA/Cloisite 30B bionanocomposites depicted in Fig. 4.15b suggested the hindrance of a dehydration process. Such a finding in thermal stability is associated with the presence of nanolayers acting as the barriers to maximise the heat insulation and minimise the permeability of volatile degradation products in the materials. This increase is also attributed to the decrease in oxygen permeability related to good clay dispersion in PVA matrices.

On the other hand, the thermal stabilities of PVA/NBC bionanocomposites are improved significantly with reference to that of PVA, as evidenced by consistently higher $T_{5\%}$, $T_{80\%}$ and T_d values, Fig. 4.14c. The degree of thermal stability of nanocomposites is even more pronounced when incorporated with NBCs in relation to $T_{5\%}$, and $T_{80\%}$. The shift of decomposition temperatures T_d and $T_d^{'}$ in Fig. 4.15c means that the dehydration process is hindered, which can result from the interaction between hydroxyl groups of PVA and hydroxyl groups on NBCs, as confirmed from previous FTIR results. Furthermore, the mass loss process occurring in the third DTG peaks means that the thermal decomposition of PVA bionanocomposites requires more reaction activation energy with a higher reaction order [55]. This finding may be attributed to the existence of NBCs working as effective barrier materials to limit the exothermicity of pyrolysis reaction with better thermal resistance of PVA bionanocomposites. The wider DTG peaks of PVA/NBC bionanocomposites beyond 3 wt% NBCs at the second decomposition step demonstrate a similar trend to those of corresponding PVA/HNT bionanocomposites along with the same dual-peak effect, as mentioned elsewhere [55].

According to the above-mentioned results, the incorporation of HNTs, Cloisite 30B clays and NBCs consistently enhances thermal properties of PVA bionanocomposites according to Table 4.1. However, the increasing rate for thermal properties has been strongly relevant to nanofiller types. It is clearly seen that the T_g of PVA bionanocomposite films with the incorporation of NBCs and Cloisite 30B clays are much higher than that of PVA/HNT bionanocomposites. Such results indicate that NBCs and Cloisite 30B clays can restrict PVA molecular chains more efficiently, as evidenced by increased mechanical properties of corresponding bionanocomposite films. According to previous studies [57, 58], the phenomenon of increasing T_g is primarily correlated to the reduction in polymeric chain mobility by incorporating inorganic nanofillers. The addition of nanoparticles into polymer matrices can change the distribution of chain segments, which is most likely due to a change in chain packing density in the vicinity of nanofiller surfaces. It should be noted that filler geometry may play a critical role in the effect of T_g. NBCs and Cloisite 30B clays have different nanofiller shapes to render the absorption of polymeric chains with the entangled structures on their surfaces when nanofiller diameters are comparable to gyration radii of polymeric chains. As such, it leads to increasing the packing density for polymeric chains and restricting their chain mobility as a result of higher T_g values. However, the incorporation of HNTs into PVA matrices has minor impact on increasing the T_g instead, which is consistent with the previous work [50]. Although the diameter of HNTs is on the nanometre scale, their length turns to be microsized, which becomes considerably higher than typical gyration radii of polymeric chains. As a consequence, it is very difficult for polymeric chains to cover entire HNT structures. Moreover, the presence of microvoids along HNT lengths could offer free sites for the segments of polymeric chains resulting in a significant increase in T_g [50].

The crystallinity of PVA bionanocomposite films reinforced with HNTs, Cloisite 30B clays and NBCs is relatively similar. Moreover, the T_m values in case of PVA/3 wt% NBC bionanocomposites and PVA/5 wt% Cloisite 30B clay bionanocomposites are nearly the same despite being slightly less than that of neat PVA at 222.91 °C. The occurrence of lower melting temperature and melting-peak broadening effects could be associated with the recrystallisation phenomenon of PVA matrices in nanocomposites [23] in that the decrease in T_m is more likely to be correlated to the changes in crystal lamellar thickness and its distribution (i.e. the formation of less perfect crystal structures).

TGA results reveal that the $T_{5\%}$ of PVA/3 wt% NBC bionanocomposites is determined to be 256.3 °C, and increases up to 262.95 °C with the incorporation of 5 wt% NBCs, which is relatively similar to that of PVA/5 wt% Cloisite 30B bionanocomposites at 261.1 °C. On the other hand, the $T_{5\%}$ of PVA/3 wt% HNT bionanocomposites appears to be determined at 265.29 °C, which is significantly higher, as compared with PVA bionanocomposites reinforced with NBCs and Cloisite 30B clays. On the contrary, the $T_{80\%}$ of PVA/3 wt% NBC bionanocomposites has been found to be 390.67 °C, and reaches 440.28 °C with the inclusion of 5 wt% NBCs, which is significantly higher than those of PVA bionanocomposites reinforced with HNTs and Cloisite 30B clays. This result suggests that the maximum thermal stability is achieved in the presence of NBCs, as opposed to HNTs and Cloisite 30B clays. In contrast to other nanofillers, the better NBC dispersion within PVA matrices takes place along with the higher barrier towards the thermal degradation. Therefore, such barrier effect can counterbalance the degradation drawback with the further improvement of thermal stability [5].

4.10 Summary

PVA bionanocomposites reinforced with different nanoparticles of Cloisite 30B clays, HNTs and NBCs were successfully prepared and characterised. The following conclusions can be drawn:

- When HNTs are embedded within PVA matrices, the properties of PVA/HNT bionanocomposites are remarkably affected and strongly depend on the HNT content. The morphological structures reveal that nanofiller dispersion can lead to various structures resulting in different enhancement levels of mechanical properties for such bionanocomposites. The incorporation of 3 wt% HNTs has improved mechanical properties while increasing the HNT content, beyond which it decreases tensile strength, elongation at break and tensile toughness of PVA/HNT bionanocomposites, resulting from debonding effect between nanofillers and polymer matrices. Moreover, thermal properties in term of degree of crystallinity, melting temperature and thermal stability of PVA/HNT bionanocomposites are enhanced, as compared with those of neat PVA films with increasing the HNT content.

- Morphological structures of PVA/Cloisite 30B clay bionanocomposites demonstrate uniform clay dispersion within PVA matrices with the formation of combined clay exfoliated and intercalated structures. These results are in good accordance with those obtained from XRD results. The tensile strengths and Young's moduli of PVA/Cloisite 30B clay bionanocomposite films are enhanced considerably with increasing the clay content up to 5 wt%, while the elongation at break and fracture toughness have been found to decrease accordingly. Thermal properties of PVA/Cloisite 30B clay bionanocomposites are improved compared to those of neat PVA films due to strong hydrogen bonding interactions between PVA matrices and nanofillers.
- The effect of different nanofiller shapes and structures on properties of PVA/NBC bionanocomposite films has also been investigated, which reveals that the maximum tensile strength and tensile modulus can be achieved with the incorporation of NBCs. This can be related to the large amount of interphase resulting from a high degree of filler dispersion in case of PVA/NBC bionanocomposites in contrast with those reinforced with HNTs and Cloisite 30B clays. Moreover, thermal stability of PVA/NBC bionanocomposites is remarkably enhanced with the inclusion of NBCs when compared to those incorporated with HNTs and Cloisite 30B clays. This can be ascribed to uniform dispersion of NBCs to generate more efficient interfacial regions than other nanofillers.

References

1. Mousa MH, Dong Y, Davies IJ (2016) Recent advances in bionanocomposites: preparation, properties, and applications. Int J Polym Mater Polym Biomater 65(5):225–254
2. Liu H, Brinson LC (2008) Reinforcing efficiency of nanoparticles: a simple comparison for polymer nanocomposites. Compos Sci Technol 68(6):1502–1512
3. Dung MX, Choi JK, Jeong HD (2013) Newly synthesized silicon quantum dot–polystyrene nanocomposite having thermally robust positive charge trapping. ACS Appl Mater Interfaces 5(7):2400–2409
4. Mousa M, Dong Y (2017) Strong poly (vinyl alcohol)(PVA)/bamboo charcoal (BC) nanocomposite films with particle size effect. ACS Sustain Chem Eng 6(1):467–479
5. Mousa M, Dong Y, Davies IJ (2018) Eco-friendly polyvinyl alcohol (PVA)/bamboo charcoal (BC) nanocomposites with superior mechanical and thermal properties. Adv Compos Mater 27(5):499–509
6. Maliakal A, Katz H, Cotts PM, Subramoney S, Mirau P (2005) Inorganic oxide core, polymer shell nanocomposite as a high K gate dielectric for flexible electronics applications. JACS 127(42):14655–14662

7. Bogdanovic U, Vodnik V, Mitric M, Dimitrijevic S, Skapin SD, Zunic V, Budimir M, Stoiljkovic M (2015) Nanomaterial with high antimicrobial efficacy-copper/polyaniline nanocomposite. ACS Appl Mater Interfaces 7(3):1955–1966
8. Mousa M, Dong Y (2018) Novel three-dimensional interphase characterisation of polymer nanocomposites using nanoscaled topography. Nanotechnology 29(38):385701
9. Komarov PV, Mikhailov IV, Chiu YT, Chen SM, Khalatur PG (2013) Molecular dynamics study of interface structure in composites comprising surface-modified SiO_2 nanoparticles and a polyimide matrix. Macromol Theory Simul 22(3):187–197
10. Meng D, Kumar SK, Lane JMD, Grest GS (2012) Effective interactions between grafted nanoparticles in a polymer matrix. Soft Matter 8(18):5002–5010
11. Ganesan V, Jayaraman A (2014) Theory and simulation studies of effective interactions, phase behavior and morphology in polymer nanocomposites. Soft Matter 10(1):13–38
12. Mousa M, Dong Y (2020) The role of nanoparticle shapes and structures in the material characterisation of polyvinyl alcohol (PVA) bionanocomposite films. Polymers 12:264
13. Ploehn HJ, Liu C (2006) Quantitative analysis of montmorillonite platelet size by atomic force microscopy. Ind Eng Chem Res 45(21):7025–7034
14. Aloui H, Khwaldia K, Hamdi M, Fortunati E, Kenny JM, Buonocore GG, Lavorgna M (2016) Synergistic effect of halloysite and cellulose nanocrystals on the functional properties of PVA based nanocomposites. ACS Sustain Chem Eng 4(3):794–800
15. Ramadan AR, Esawi AM, Gawad AA (2010) Effect of ball milling on the structure of Na^+-montmorillonite and organo-montmorillonite (Cloisite 30B). Appl Clay Sci 47(3–4):196–202
16. Mallakpour S, Shahangi V (2013) Bio-modification of Cloisite Na^+ with chiral L-leucine and preparation of new poly (vinyl alcohol)/organo-nanoclay bionanocomposite films. Synth React Inorg M 43(8):966–971
17. Fu D, Zhang Y, Lv F, Chu PK, Shang J (2012) Removal of organic materials from TNT red water by Bamboo Charcoal adsorption. Chem Eng J 193:39–49
18. She B, Tao X, Huang T, Lu G, Zhou Z, Guo C, Dang Z (2016) Effects of nano bamboo charcoal on PAHs-degrading strain *Sphingomonas* sp. GY2B. Ecotoxicol Environ Saf 125:35–42
19. Lorenzoni M, Evangelio L, Verhaeghe S, Nicolet C, Navarro C, Pérez-Murano F (2015) Assessing the local nanomechanical properties of self-assembled block copolymer thin films by peak force tapping. Langmuir 31(42):11630–11638
20. Babu SS, Mathew S, Kalarikkal N, Thomas S (2016) Antimicrobial, antibiofilm, and microbial barrier properties of poly (ε-caprolactone)/Cloisite 30B thin films. 3 Biotech 6(2):249
21. Huang D, Wang W, Kang Y, Wang A (2012) A chitosan/poly (vinyl alcohol) nanocomposite film reinforced with natural halloysite nanotubes. Polym Compos 33(10):1693–1699
22. Abdullah ZW, Dong Y, Davies IJ, Barbhuiya S (2017) PVA, PVA blends, and their nanocomposites for biodegradable packaging application. Polym Plast Technol Eng 56(12):1307–1344
23. Carli LN, Crespo JS, Mauler RS (2011) PHBV nanocomposites based on organomodified montmorillonite and halloysite: the effect of clay type on the morphology and thermal and mechanical properties. Compos Part A Appl Sci Manuf 42:1601–1608
24. Ray SS, Okamoto M (2003) Polymer/layered silicate nanocomposites: a review from preparation to processing. Prog Polym Sci 28(11):1539–1641
25. Swapna VP, Suresh KI, Saranya V, Rahana MP, Stephen R (2015) Thermal properties of poly (vinyl alcohol)(PVA)/halloysite nanotubes reinforced nanocomposites. Int J Plast Technol 19(1):124–136
26. Rhim JW, Hong SI, Ha CS (2009) Tensile, water vapor barrier and antimicrobial properties of PLA/nanoclay composite films. LWT-Food Sci Technol 42(2):612–617
27. Strawhecker KE, Manias E (2000) Structure and properties of poly (vinyl alcohol)/Na+ montmorillonite nanocomposites. Chem Mater 12(10):2943–2949
28. Gaume J, Taviot-Gueho C, Cros S, Rivaton A, Therias S, Gardette JL (2012) Optimization of PVA clay nanocomposite for ultra-barrier multilayer encapsulation of organic solar cells. Sol Energy Mater Sol Cells 99:240–249
29. Li S, Li X, Deng Q, Li D (2015) Three kinds of charcoal powder reinforced ultra-high molecular weight polyethylene composites with excellent mechanical and electrical properties. Mater Des 85:54–59

30. Liang J, Huang Y, Zhang L, Wang Y, Ma Y, Guo T, Chen Y (2009) Molecular-level dispersion of graphene into poly (vinyl alcohol) and effective reinforcement of their nanocomposites. Adv Funct Mater 19(14):2297–2302
31. Cheng ZL, Qin XX, Liu Z, Qin DZ (2017) Electrospinning preparation and mechanical properties of PVA/HNTs composite nanofibers. Polym Adv Technol 28(6):768–774
32. Azmi S, Razak SIA, Abdul Kadir MR, Iqbal N, Hassan R, Nayan NHM, Abdul Wahab AH, Shaharuddin S (2017) Reinforcement of poly (vinyl alcohol) hydrogel with halloysite nanotubes as potential biomedical materials. Soft Mater 15(1):45–54
33. Ormanci-Acar T, Celebi F, Keskin B, Mutlu-Salmanlı O, Agtas M, Turken T, Tufani A, Imer DY, Ince GO, Demir TU, Menceloglu YZ (2018) Fabrication and characterization of temperature and pH resistant thin film nanocomposite membranes embedded with halloysite nanotubes for dye rejection. Desalination 429:20–32
34. Hakalahti M, Mautner A, Johansson LS, Hänninen T, Setälä H, Kontturi E, Bismarck A, Tammelin T (2016) Direct interfacial modification of nanocellulose films for thermoresponsive membrane templates. ACS Appl Mater Interfaces 8(5):2923–2927
35. Humood M, Qin S, Song Y, Polychronopoulou K, Zhang Y, Grunlan JC, Polycarpou AA (2016) Influence of graphene reduction and polymer cross-linking on improving the interfacial properties of multilayer thin films. ACS Appl Mater Interfaces 9(1):1107–1118
36. Carrado KA, Thiyagarajan P, Elder DL (1996) Polyvinyl alcohol-clay complexes formed by direct synthesis. Clays Clay Miner 44(4):506–514
37. Theng BKG (1982) Clay–polymer interactions: summary and perspectives. Clays Clay Miner 30(1):1–10
38. Alishahi E, Shadlou S, Doagou-R S, Ayatollahi MR (2013) Effects of carbon nanoreinforcements of different shapes on the mechanical properties of epoxy-based nanocomposites. Macromol Mater Eng 298(6):670–678
39. Abdullah ZW, Dong Y (2018) Preparation and characterisation of poly (vinyl) alcohol (PVA)/starch (ST)/halloysite nanotube (HNT) nanocomposite films as renewable materials. J Mater Sci 53(5):3455–3469
40. Khoo WS, Ismail H, Ariffin A (2013) Tensile, swelling, and oxidative degradation properties of crosslinked polyvinyl alcohol/chitosan/halloysite nanotube composites. Int J Polym Mater Polym Biomater 62(7):390–396
41. Qiu K, Netravali AN (2013) Halloysite nanotube reinforced biodegradable nanocomposites using noncrosslinked and malonic acid crosslinked polyvinyl alcohol. Polym Compos 34(5):799–809
42. Sapalidis AA, Katsaros FK, Kanellopoulos NK (2011) PVA/montmorillonite nanocomposites: development and properties. In: Cuppoletti J (ed) Nanocomposites and polymers with analytical methods. InTech, Rijeka, pp 29–50
43. Asad M, Saba N, Asiri AM, Jawaid M, Indarti E, Wanrosli WD (2018) Preparation and characterization of nanocomposite films from oil palm pulp nanocellulose/poly (vinyl alcohol) by casting method. Carbohydr Polym 191:103–111
44. Raheel M, Yao K, Gong J, Chen XC, Liu DT, Lin YC, Cui DM, Siddiq M, Tang T (2015) Poly (vinyl alcohol)/GO-MMT nanocomposites: preparation, structure and properties. Chin J Polym Sci 33(2):329–338
45. Schadler LS, Brinson LC, Sawyer WG (2007) Polymer nanocomposites: a small part of the story. JOM 59(3):53–60
46. Pakzad A, Simonsen J, Yassar RS (2012) Gradient of nanomechanical properties in the interphase of cellulose nanocrystal composites. Compos Sci Technol 72(2):314–319
47. Santos KS, Castel CD, Liberman SA, Oviedo MAS, Mauler RS (2011) Polyolefin-based nanocomposite: the effects of processing aids. J Appl Polym Sci 119(3):1567–1575
48. Wetzel B, Rosso P, Haupert F, Friedrich K (2006) Epoxy nanocomposites–fracture and toughening mechanisms. Eng Fract Mech 73(16):2375–2398
49. Li Y, Yang T, Yu T, Zheng L, Liao K (2011) Synergistic effect of hybrid carbon nantube–graphene oxide as a nanofiller in enhancing the mechanical properties of PVA composites. J Mater Chem 21(29):10844–10851

50. Liu M, Guo B, Du M, Jia D (2007) Drying induced aggregation of halloysite nanotubes in polyvinyl alcohol/halloysite nanotubes solution and its effect on properties of composite film. Appl Phys A 88(2):391–395
51. Yu Y-H, Lin CY, Yeh JM, Lin WH (2003) Preparation and properties of poly(vinyl alcohol)-clay nanocomposite materials. Polymer 44:3553–3560
52. Sirousazar M, Kokabi M, Hassan ZM, Bahramian AR (2012) Polyvinyl alcohol/Na-montmorillonite nanocomposite hydrogels prepared by freezing-thawing method: structural, mechanical, thermal, and swelling properties. J Macromol Sci Part B 51:1335–1350
53. Jose T, George SC, Maria HJ, Wilson R, Thomas S (2014) Effect of bentonite clay on the mechanical, thermal, and pervaporation performance of the poly(vinyl alcohol) nanocomposite membranes. Ind Eng Chem Res 53:16820–16831
54. Ali SS, Tang X, Alavi S, Faubion J (2011) Structure and physical properties of starch/poly vinyl alcohol/sodium montmorillonite nanocomposite films. J Agr Food Chem 59:12384–12395
55. Peng Z, Kong LX (2007) A thermal degradation mechanism of polyvinyl alcohol/silica nanocomposites. Polym Degrad Stab 92(6):1061–1071
56. Mondal D, Mollick MMR, Bhowmick B, Maity D, Bain MK, Rana D, Mukhopadhyay A, Dana K, Chattopadhyay D (2013) Effect of poly (vinyl pyrrolidone) on the morphology and physical properties of poly (vinyl alcohol)/sodium montmorillonite nanocomposite films. Prog Nat Sci Mater Int 23(6):579–587
57. Nakane K, Yamashita T, Iwakura K, Suzuki F (1999) Properties and structure of poly (vinyl alcohol)/silica composites. J Appl Polym Sci 74(1):133–138
58. Mbhele ZH, Salemane MG, Van Sittert CGCE, Nedeljković JM, Djoković V, Luyt AS (2003) Fabrication and characterization of silver–polyvinyl alcohol nanocomposites. Chem Mater 15(26):5019–5024

Chapter 5
3D Interphase of PVA Bionanocomposite Films

Abstract In this chapter, nanomechanical properties and interphase dimensions of polyvinyl alcohol (PVA) bionanocomposite films were measured by a peak force quantitative nanomechanical tapping mode (PFQNM). Our results revealed that the interphase in case of PVA/nanodiameter bamboo charcoal (NBC) bionanocomposites has higher nanoelastic properties and reinforcement efficiency than those of PVA/halloysite nanotube (HNT) bionanocomposites and PVA/Cloisite 30B clay bionanocomposites. Such results are associated with higher interphase volume per unit particle volume in case of PVA/NBC bionanocomposites, as opposed to the other two bionanocomposites.

Keywords 3D interphase · Bionanocomposites · Nanomechanical properties · Analytical modelling

5.1 Introduction

Recently, it has been proven that the formation of interphase region in the vicinity of nanofillers induces the dramatic change of properties of polymer nanocomposites in contrast to those of corresponding polymer matrices [1]. Such an interphase region can be told from nanoparticles and polymer matrices owing to their distinct physical properties in that material performance of nanocomposite materials primarily relies on the volume and properties of interfacial regions [2, 3]. Different factors such as particle structure and shape can significantly affect the interaction level between particles and polymeric molecules, which plays an important role in determining interphase dimensions and properties. This is particularly the case in this study where the incorporation of layered montmorillonite (MMT) with PVA can result in different matrix–particle interactions and diverse interphase properties when compared with the use of tubular HNTs and NBCs, as reported in the following sections. Theoretical simulations, such as atomistic and coarse-grained molecular

This chapter takes partial content materials from authors' published research article **'Mousa M, Dong Y (2020) Towards sophisticated 3D interphase modelling of advanced bionanocomposites via atomic force microscopy, J Nanomater, Article ID 4526108'** under Creative Commons Attribution Licence.

M. Mousa and Y. Dong, *Multiscaled PVA Bionanocomposite Films*,
https://doi.org/10.1007/978-981-15-8771-9_5

simulations, have made significant progress in evaluating polymer–particle interactions. However, with the limited computational environment required, these studies have been mostly restricted to the context of single and two-particle systems. The peak force quantitative nanomechanical mapping (PFQNM) becomes a relatively new and powerful technique, which was quantitatively utilised to acquired interphase dimensions and properties based on the variation in nanomechanical properties of materials such as stiffness and adhesion of nanocomposites along with corresponding acquired dimensions [3].

5.2 Interphase Characterisation of PVA-Based Bionanocomposites

5.2.1 *Modelling Approach*

5.2.1.1 Interphase Modulus

The elastic properties of interphases such as interphase modulus between PVA matrices and dispersed anisotropic NBCs in various shapes and sizes were determined according to a data set of elastic moduli collected based on PVA/NBC interphases surrounding 75 different NBCs at 25 line scan regions (LSRs). The same procedure was employed to assess interphase moduli of PVA/HNT bionanocomposites and PVA/Cloisite 30B clay bionanocomposites, respectively.

5.2.1.2 Interphase Dimensions

Interphase dimensions in terms of interphase width $W_{i\ \mathrm{Interphase}}$, interphase length $L_{j\ \mathrm{Interphase}}$ and interphase height $H_{k\ \mathrm{Interphase}}$ were measured by scanning individual PVA/nanoparticle phases using PFQNM with typical features of distinct interphases between nanoparticles and polymer matrices based on the variation of their nanomechanical properties. For instance, $W_{i\ \mathrm{Interphase}}$ was measured by scanning along the *i*th transverse plane (i = 1, 2, 3, …) for PVA/nanoparticle phases. The same procedure reapplied to determine $L_{j\mathrm{Interphase}}$ and $H_{k\ \mathrm{Interphase}}$ along the *j*th longitudinal plane (j = 1, 2, 3, …) and *k*th height plane (k = 1,2, 3, …), respectively, (refer to Appendix for more details).

On the other hand, it was assumed that uniform nanoparticles dispersion took place with two typical categories, namely fully embedded and partially embedded NBCs, HNTs and Cloisite 30B clays within PVA matrices in resulting bionanocomposites, as illustrated in Fig. 5.1. It is clearly demonstrated that the interphase is surrounded between inner interface area and outer interface area bound by nanofillers and PVA matrices, respectively. By rearranging analytical equations according to Behmer and Hawkins [4] used for calculating the surface areas of anistropic shapes,

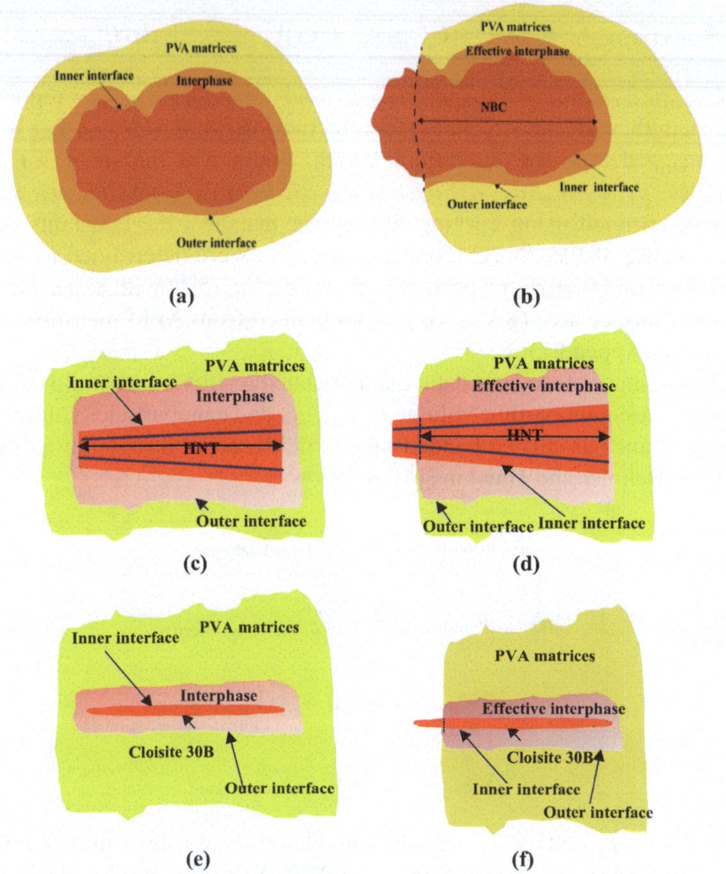

Fig. 5.1 Schematic diagrams of fully embedded nanoparticles: **a** NBCs, **c** HNTs and **e** Cloisite 30B clays, as well as partially embedded nanoparticles: **b** NBCs, **d** HNTs and **f** Cloisite 30B clays in PVA bionanocomposite systems [3, 5]

the following Eqs. (5.1)–(5.4) are derived for determining both surface areas of outer interface ($SA_{\text{outer Interface}}$) and inner interface ($SA_{\text{inner Interface}}$) in a wide range of fully and partially embedded nanoparticles. It is noted that subscripts of 'f' and 'p' mean fully and partially embedded nanoparticles, respectively, as depicted in Fig. 5.1.

$$(SA_{\text{outer Interface}})_f = a_1 + b_1 L^2_{\text{Interphase}} + c_1 W^2_{\text{Interphase}} + d_1 H^2_{\text{Interphase}} \quad (5.1)$$

$$(SA_{\text{outer Interface}})_p = a_2 + b_2 L'^2_{\text{Interphase-effective}} + c_2 W'^2_{\text{Interphase-effective}} + d_2 H'^2_{\text{Interphase-effective}} \quad (5.2)$$

$$(SA_{\text{inner Interface}})_f = a_3 + b_3 L^2_{\text{p}} + c_3 W^2_{\text{P}} + d_3 H^2_{\text{P}} \quad (5.3)$$

$$(SA_{\text{inner Interface}})_p = a_4 + b_4 L'^2_{\text{p-effective}} + c_4 W'^2_{\text{p-effective}} + d_4 H'^2_{\text{p-effective}} \quad (5.4)$$

where $L'_{\text{Interphase-effective}}$, $W'_{\text{Interphase-effective}}$ and $H'_{\text{Interphase-effective}}$ represent the maximum length, width and thickness for effective interphases. $L'_{\text{p-effective}}$, $W'_{\text{p-effective}}$ and $H'_{\text{p-effective}}$ denote the maximum length, width and thickness for effective nanoparticles. a_1, b_1, c_1, d_1, a_2, b_2, c_2, d_2, a_3, b_3, c_3, d_3 and a_4, b_4, c_4, d_4 are constants determined using curving fitting. For instance, the constants of a_1, b_1, c_1 and d_1 in case of PVA/NBC bionanocomposites were determined to be 0.7439, 0.3627, 0.7006 and 0.9979, respectively, by fitting Eq. (5.1) with experimental data in relation to surface area $(SA'_{\text{outer Interface}})_f$ obtained from AFM measurements (see more details in Appendix).

After obtaining surface area data, the associated results were employed in determining interphase/nanoparticle volume $V_{\text{p/Interphase}}$ and nanoparticles volume V_p with respect to fully and partially embedded nanoparticles according to modified equations derived from Behmer and Hawkins [4] as follows:

$$(SA_{\text{outer Interface}})_f = e_1 V^{f_1}_{\text{p/Interphase}} \quad (5.5)$$

$$(SA_{\text{outer Interface}})_p = e_2 V^{f_2}_{\text{p/Interphase effective}} \quad (5.6)$$

$$(SA_{\text{inner Interface}})_f = e_3 V^{f_3}_{p} \quad (5.7)$$

$$(SA_{\text{inner Interface}})_p = e_4 V^{f_4}_{\text{p effective}} \quad (5.8)$$

$e_1, f_1, e_2, f_2, e_3, f_3$ and e_4, f_4 are constants determined using curving fitting. For example, the constants of e_1 and f_1 in case of PVA/NBC phases were estimated to be 0.3824 and 0.3825, respectively, by fitting Eq. (5.5) with the data value of $V_{\text{p/Interphase}}$ obtained from the AFM measurement (refer to Appendix for more details).

Final step is to determine interphase volume $V_{\text{Interphase}}$ for fully or partially embedded nanoparticles in PVA nanocomposites shown in Figs. 5.1 as follows:

$$\left(V_{\text{Interphase}}\right)_f = V_{\text{p/Interphase}} - V_p \quad (5.9)$$

$$\left(V_{\text{Interphase}}\right)_p = V_{\text{p/Interphase-effective}} - V_{\text{p effective}} \quad (5.10)$$

Note that the aforementioned procedure applied to three PVA bionanocomposite systems reinforced with NBCs, HNTs and Cloisite 30B clays used in this study by considering interphase dimensions and properties (refer to Appendix for more details).

5.2.2 Modulus-Gradient Effect

As illustrated in Fig. 5.2a, strong interactions between NBCs and PVA matrices appear to be not only on NBC surfaces like other nanoparticles such as CNTs and nanosillca but also into entire NBC particles through the penetration of PVA molecules. Figure 5.2b, c demonstrates a typical topographical image of PVA/NBC

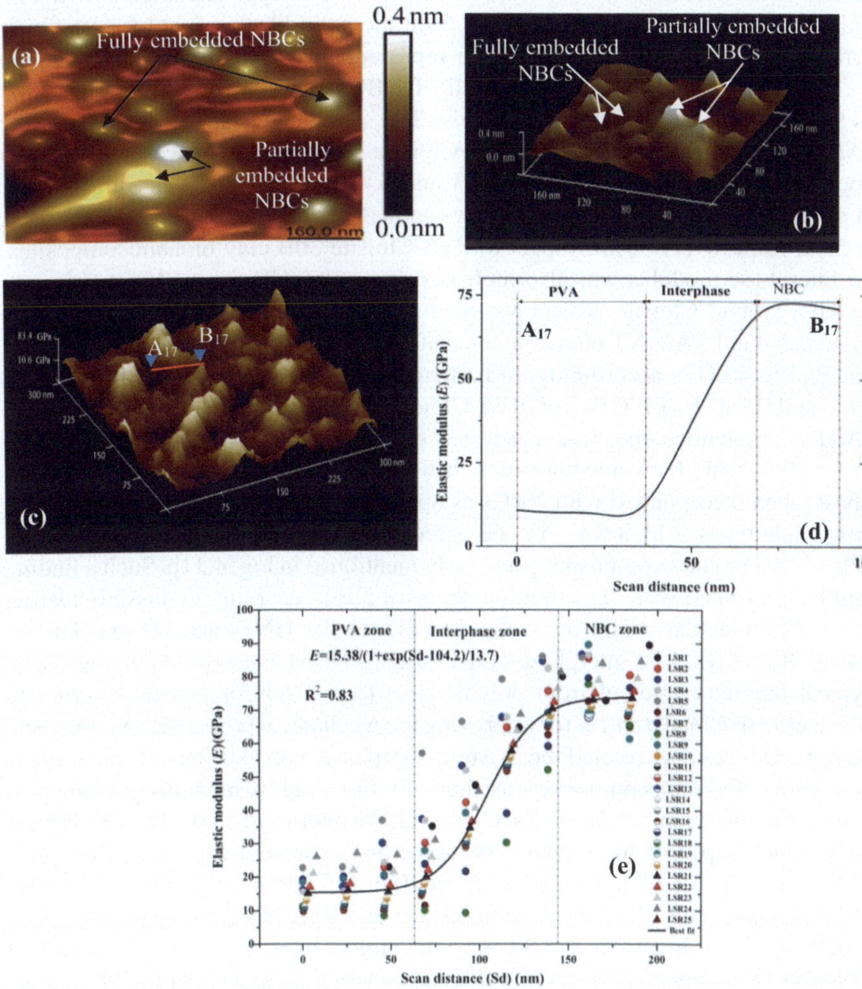

Fig. 5.2 Proposed schemes for typical PVA/NBC interaction with fully and partially embedded NBCs in **a** 2D view and **b** 3D view, **c** 3D AFM modulus mapping image of PVA/3 wt% NBC bionanocomposites, **d** modulus profile for PVA/3 wt% NBC bionanocomposites taken along the cut section $A_{17}B_{17}$ and **e** typical data sets of modulus profiles along 25 line scan regions (LSRs) of corresponding bionanocomposites with the best-fit curve in which S_d represents scan distance [3]

nanocomposite sample and the variation of elastic moduli from PVA matrices, interphases to NBCs, respectively. According to a typical cut line $A_{17}B_{17}$ shown in Fig. 5.2c, elastic modulus has significantly increased from the lowest modulus level of 14.8 GPa with respect to PVA matrices to the highest of 72.86 GPa for NBCs. In particular, the interphase exhibits an almost linearly increasing relationship of its elastic modulus with scan distance except interphase boundary regions around PVA matrices and NBCs alone, which is in a range from 17.1 GPa near the PVA region to 64.9 GPa around NBCs with the measured interlayer thickness of 31.8 nm. Final interphase modulus was obtained from the solid black curve by best fitting discrete modulus data sets based on 25 representative LSRs in Fig. 5.2d, which also applied to determine elastic moduli of NBCs and PVA matrices to be 78.4 $\pm$ 4.9 and 24.25 $\pm$ 4.2 GPa, respectively. The same analysis was carried out for PVA/HNT bionanocomposites and PVA/Cloisite 30B clay bionanocomposites with associated results presented in Figs. 5.3 and 5.4. It was found that similar fully and partially embedded Cloisite 30B clays and HNTs were manifested in Figs. 5.3a and 5.4a, respectively. With respect to PVA/Cloisite 30B clay bionanocomposites, their interphase modulus was also increased from 20.6 GPa near PVA matrices to 42.2 GPa around Cloisite 30B clays. Such a modulus-gradient phenomenon was also revealed in PVA/HNT bionanocomposites with the interphase modulus varying from 19.3 to 40 GPa accordingly. In particular, PVA modulus appeared to be 21.4 $\pm$ 4.1 and 19.8 $\pm$ 3.7 GPa for PVA/Cloisite 30B clay bionanocomposites and PVA/HNT bionanocomposites, which was equivalent to 13.4 and 22.5% less than that for PVA/NBC bionanocomposites. It was implied that PVA modulus was the highest when incorporated with NBCs as opposed to the inclusion of the other two nanoparticle types, which was in good agreement with higher tensile strength data of PVA/NBC bionanocomposites previously mentioned in Fig. 4.11b. Such a finding could be associated with 3D nanofiller shape of NBCs enabling to possibly restrict more PVA molecular chains as opposed to 2D tubular HNTs and 1D platelet-like Cloisite 30B clays. The interphase could be considered as transitional zones with a typical modulus gradient from polymer matrices to reinforcements in order to achieve effective load transfer for improving the mechanical properties of composite materials. Our results revealed the existing interphase with excellent elastic properties in PVA/NBC nanocomposites, as shown in Fig. 5.2d. A modulus gradient was detected ranging from 25.32 $\pm$ 3.4 GPa for PVA matrices to 66.3 $\pm$ 3.2 GPa for NBCs, which appeared to be relatively high when compared with those from 22.2 $\pm$ 2.9 to 43.4 $\pm$ 3.1 GPa, as well as from 20.6 $\pm$ 3.4 to 41.6 $\pm$ 4.6 GPa in case of PVA/Cloisite 30B clay bionanocomposites and PVA/HNT bionanocomposites, respectively, according to Figs. 5.3d and 5.4d. Higher interphase modulus identified in PVA/NBC bionanocomposites may directly benefit from highly porous NBC structures to cover all their 3D structures when compared with 1D platelet-like Cloisite 30B clays and 2D tubular HNTs as reinforcements. This can further facilitate inducing a highly positive capillary pressure to drive PVA molecular chains into NBC pores with typical resulting chemical bonding in bionanocomposites [2, 3]. These NBC pores also possess 'mechanical anchoring' mechanism [6] when interacting with PVA molecular chains, which means that a mechanical interlocking phenomenon

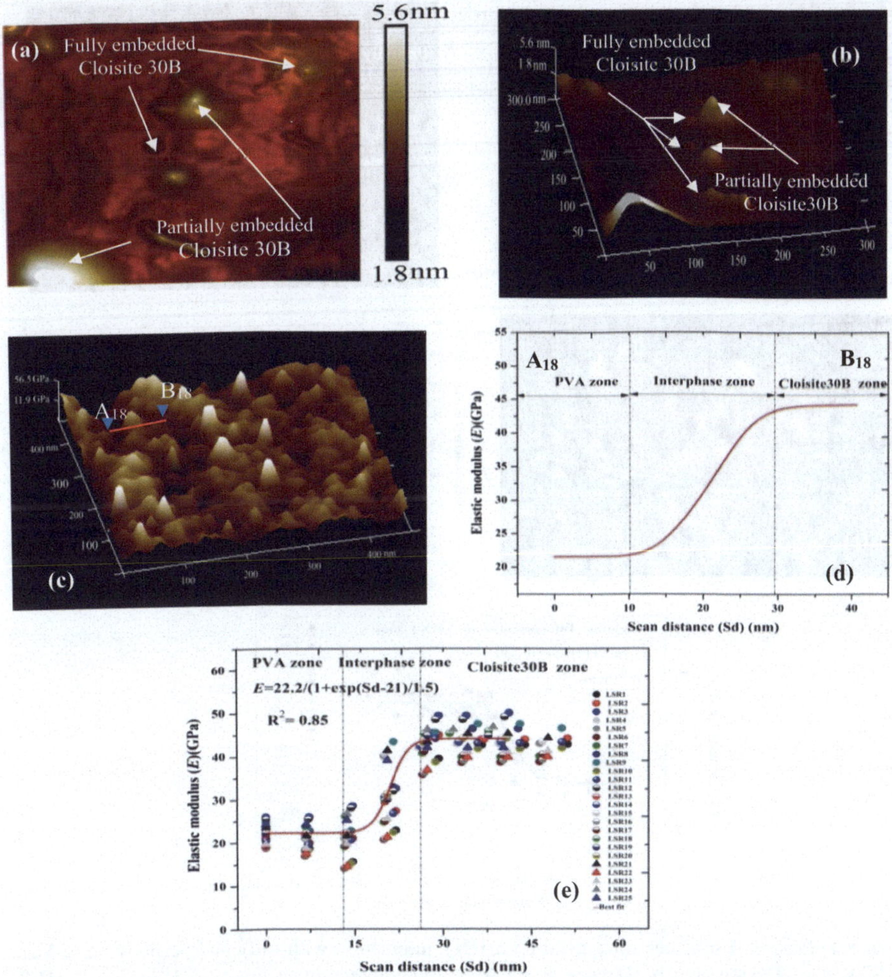

Fig. 5.3 Proposed schemes for typical PVA/Cloisite 30B clay interaction with fully and partially embedded clays in **a** 2D view and **b** 3D view, **b** 3D AFM modulus mapping image of PVA/3 wt% Cloisite 30B clay bionanocomposites, **c** modulus profile for PVA/3 wt% Cloisite 30B clay bionanocomposite along the cut section $A_{18}B_{18}$ [5] and **d** typical data sets of modulus profiles along 25 line scan regions (LSRs) of corresponding bionanocomposites with the best-fit curve in which S_d represents scan distance [5]

takes place due to the existence of PVA molecular chains inside BC surfaces [6]. On the contrary, the interactions of Cloisite 30B clays and HNTs within PVA matrices are limited to the creation of hydrogen bonding between PVA molecular chains and those detected on their nanoparticle surfaces. Besides, the formation of mechanical bonding from Cloisite 30B clays or HNTs to PVA matrices is very limited. Moreover, the modulus-gradient effect at the interphase zones between nanofillers

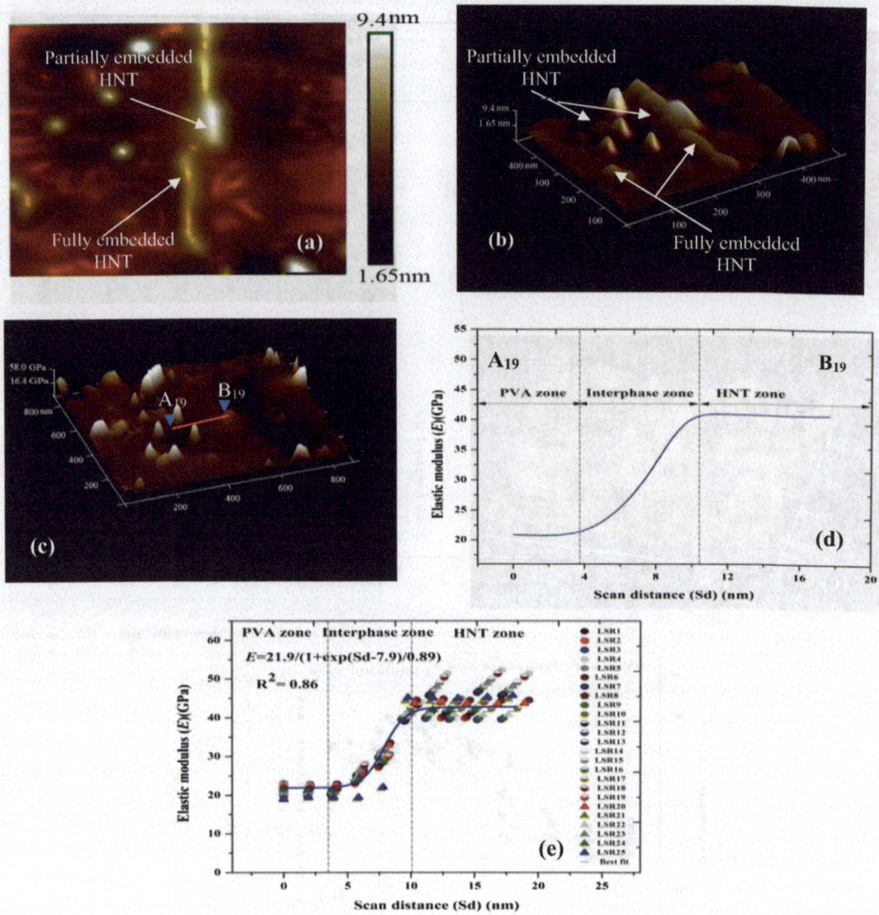

Fig. 5.4 Proposed schemes for typical PVA/HNT interaction with fully and partially embedded HNTs in **a** 2D view and **b** 3D view, **c** 3D AFM modulus mapping image of PVA/3 wt% HNT bionanocomposites [5], **d** modulus profile for PVA/3 wt% HNT bionanocomposites taken along the cut section $A_{19}B_{19}$ [5] and **e** typical data sets of modulus profiles along 25 line scan regions (LSRs) of corresponding bionanocomposites with the best-fit curve in which S_d represents scan distance

and PVA matrices is associated with the gradient number of hydrogen bonding from nanofillers surfaces to polymer matrices, which plays an important role in controlling mechanical properties of polymer nanocomposites [7, 8]. Nanoelastic behaviour of PVA demonstrates an elastic modulus range of 19.8–24.25 GPa in this study, which is close to the value of 23.69 GPa previously reported via the nanoidentation of PVA/chitsan (CS) copolymer coating [9]. However, the average elastic modulus of bulk PVA films was approximately 2.08 GPa at a macroscopic level, which appeared to be much smaller in contrast with those of individual amorphous phase at 11.4 ± 3.1 GPa and crystalline phase at 24 ± 4.2 GPa, which were obtained via PFQNM for PVA/NBC bionanocomposites shown in Fig. 3.10c. Such a great difference in measurements can be ascribed to the following reasons: (i) Though instrumental parameters have been calibrated prior to AFM measurements, the shape function of the topmost probe tip may not be accurate enough for the low penetration depth [10, 11]. In this study, the low penetration force has been used, resulting in a low penetration depth in order to minimise the effects of residual stresses, as well as plastic deformation from the neighbouring indent with the capability of high lateral resolution particularly used for nanointerphases. (ii) Nanosurface properties often vary from those bulk properties because of the discrepancy in morphological structures between outer skins on material surfaces and bulk films, as well as the differences in nanomechanical behaviour and bulk properties [11]. (iii) Despite the data reliability and repeatability of DMT modulus, its determination using the PFQNM is apparently distinct from tensile modulus for bulk properties using conventional tensile testing due to different measurement mechanisms [12].

5.2.3 3D Interphase Dimensions and Modulus

Elastic properties of interphase generally vary when surrounded by individual particles in different sizes and shapes, thus leading to the change of interphase dimensions. Interphase dimensions can be determined according to AFM height and adhesion images of PVA/NBC bionanocomposite samples, depicted in Fig. 5.5a, b, as well as Fig. 5.5c, d, respectively. As evidently seen in Fig. 5.5a, b, the height of NBCs is generally much greater than that of PVA. It is clearly shown that a linearly increasing tendency in height occurs from interphase regions near PVA matrices to those surrounding NBCs. NBCs tend to be easily distinguished from PVA matrices due to their different adhesion properties. The tip adhesion to PVA matrices in PVA/NBC bionanocomposites was determined to be 10.76 $\pm$ 3.42 nN, which was found to be over five-fold greater than that of NBCs at 2.1 $\pm$ 0.87 nN. Such a finding can be associated with the hydrophilic nature of PVA with relatively high adhesion force when compared with hydrophobic NBCs. Accordingly, interphase thickness is represented by the scan distances of 16 and 13 nm with a sharp decreasing adhesion gradient from PVA matrices to NBCs on both sides of selected material regions, as illustrated in Fig. 5.5d. This result is in good accordance with the values of 13 and 12.5 nm, previously determined for epoxy/graphene nanoplatelet (GNP) composites and epoxy/graphene oxide (GO) nanocomposites, respectively [13]. A similar procedure was used to detect interphase thickness in PVA/HNT bionanocomposites and PVA/Cloisite 30B clay bionanocomposites according to height and adhesion profiles depicted in Figs. 5.6 and 5.7, respectively. In both bionanocomposite systems, the tip adhesion to PVA matrices was 14.24 $\pm$ 2.31 nN, which appeared to be greater than those for HNTs and Cloisite 30B clays at 8.6 $\pm$ 1.08 and 8.4 $\pm$ 1.1 nN, respectively. It was manifested that the tip adhesions to HNTs and Cloisite 30B clays were still much higher than that of NBCs at 2.1 $\pm$ 0.87 nN, resulting from more hydroxyl groups detected in HNTs and Cloisite 30B clays according to previous FTIR results

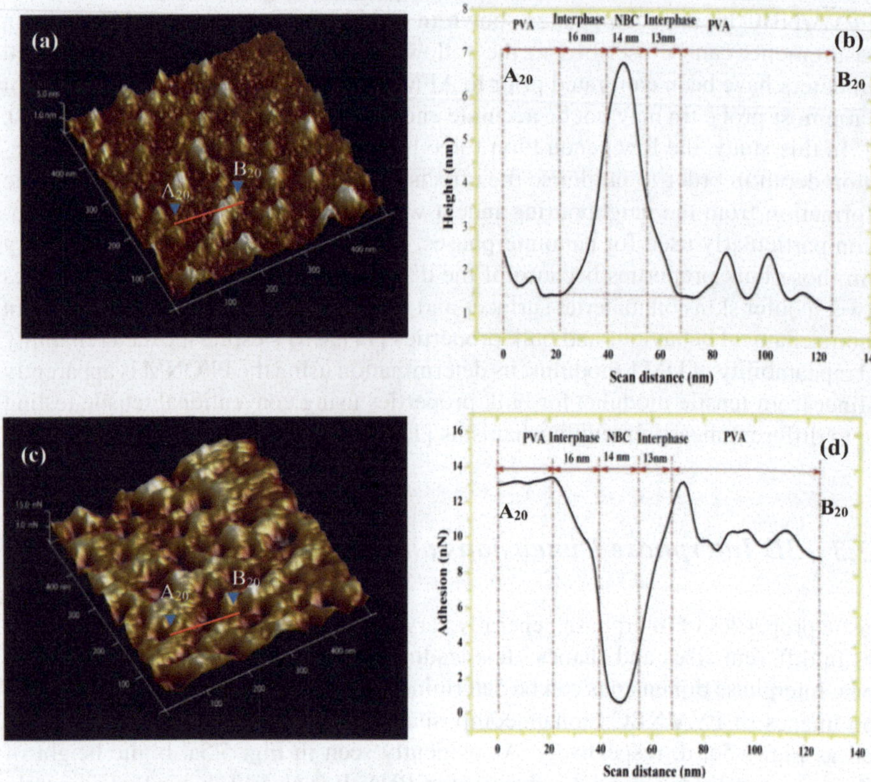

Fig. 5.5 **a** 3D AFM height mapping image of PVA/3 wt% NBC bionanocomposites, **b** height profile of the sample taken along the cut section $A_{20}B_{20}$, **c** 3D AFM adhesion mapping image of PVA/3 wt% NBC bionanocomposites and **d** adhesion profile of corresponding bionanocomposite sample taken along the cut section $A_{20}B_{20}$ [3]

in Fig. 4.2a, b. As illustrated in Fig. 5.6d, interphase thickness was found to be 6.5 and 9.2 nm on both sides of HNTs when compared with 8.4 and 10.2 nm on those of Cloisite 30B clays in Fig. 5.7d. In comparison, larger interphase thickness around NBCs was revealed relative to those in the vicinity of HNTs and Cloisite 30B clays, resulting in stronger bonding effect between nanofillers and polymer matrices. Consequently, it is proven from our results that interphase thickness $t_{\mathrm{Interphase}}$ is a nonuniform and non-constant quantity as far as interphase dimensions are concerned. The uniformity of $t_{\mathrm{Interphase}}$ is most likely to be associated with the number of chemical hydrogen bonds and physical roughness of NBC surfaces [14, 15]. The major drawback in previous work lies in a simple assumption of one-dimensional interphase layers with constant interphase thickness. However, our study demonstrates that actual nanointerphases should be investigated from a three-dimensional point of view to identify the nonuniformity and dimensional variations of interphases in terms of $t_{\mathrm{Interphase}}$. Here, we successfully identified interphase dimensions in term

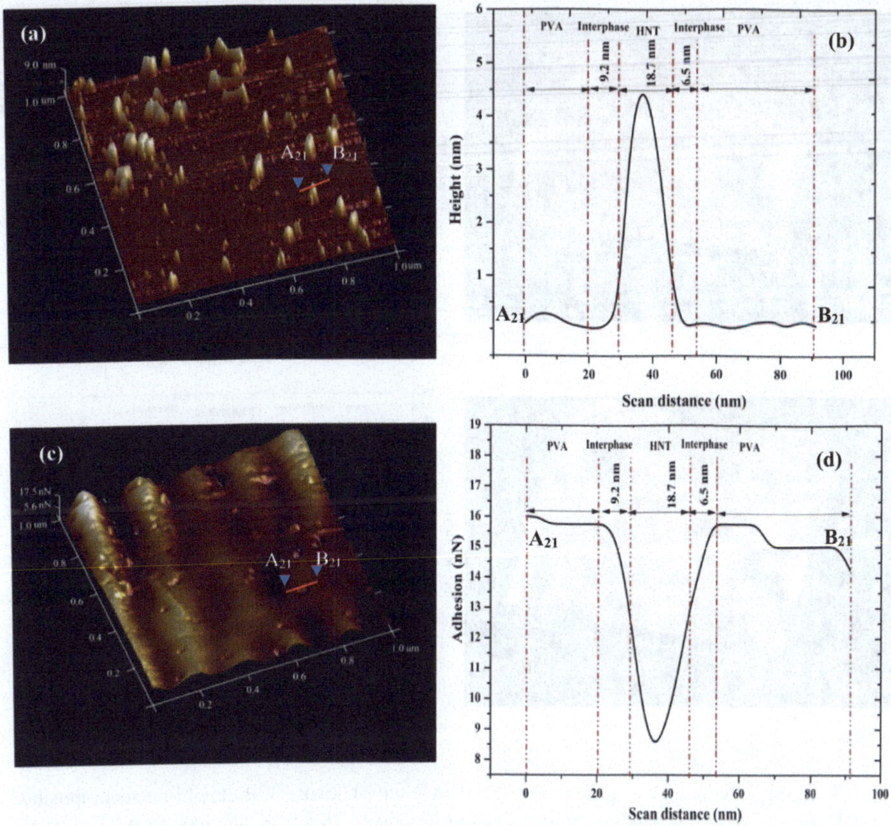

Fig. 5.6 **a** 3D AFM height mapping image of PVA/3 wt% HNT bionanocomposites, **b** height profile of the sample taken along the cut section $A_{21}B_{21}$, **c** 3D AFM adhesion mapping image of PVA/3 wt% HNT bionanocomposites and **d** adhesion profile of corresponding bionanocomposite sample taken along the cut section $A_{21}B_{21}$ [5]

of interphase width ($W_{\text{Interphase}}$), interphase length ($L_{j\text{Interphase}}$) and interphase height ($H_{k\text{Interphase}}$) among individual NBCs, as shown in Fig. 5.8a. It is clearly seen that the parameters of $L_{\text{Interphase}}$, $W_{\text{Interphase}}$ and $H_{\text{Interphase}}$ were measured to be 97, 40 and 8 nm, respectively. Moreover, according to Fig. 5.8b, it appears that there is no direct relationship between interphase thickness and NBC thickness, suggesting that interphase thickness may be independent of particle thickness. On the other hand, it is evidently seen from Fig. 5.8b–e that $H_{\text{Interphase}}$ tends to decrease in a linear manner with an increase in H_{NBC}, while $L_{\text{Interphase}}$ and $W_{\text{Interphase}}$ linearly increase with increasing L_{NBC} and W_{NBC}, respectively. This phenomenon is ascribed to more contact surface areas taking place between polymer matrices and nanoparticles in a nanocomposite system [16].

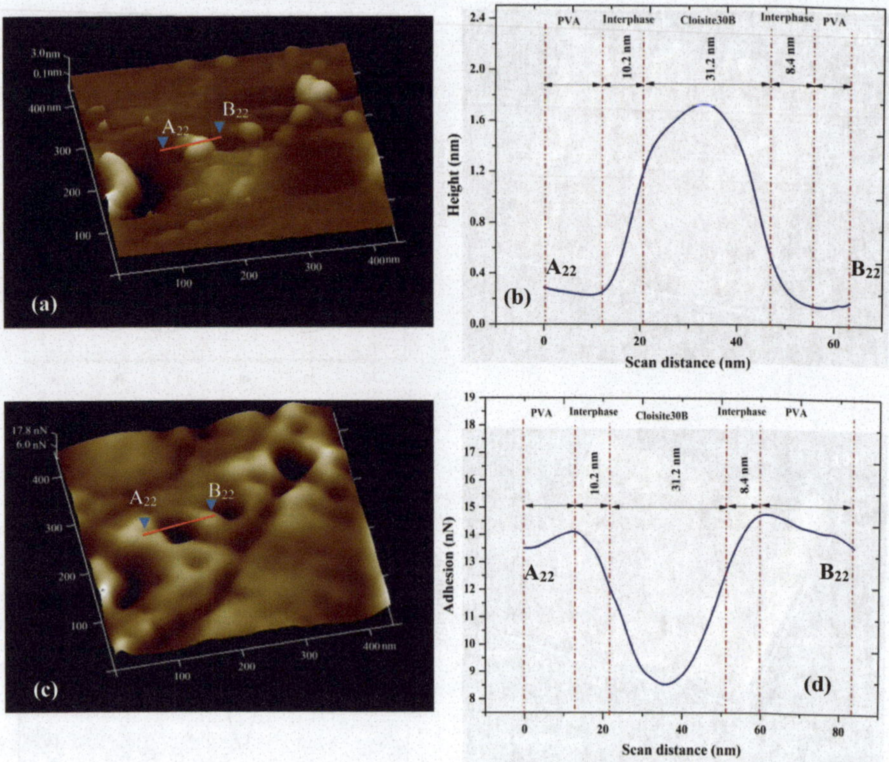

Fig. 5.7 **a** 3D AFM height mapping image of PVA/3 wt% Cloisite 30B clay bionanocomposites, **b** height profile of corresponding bionanocomposites taken along the cut section $A_{22}B_{22}$, **c** 3D adhesion mapping image of PVA/3 wt% Cloisite 30B clay bionanocomposites and **d** their adhesion profile taken along the cut section $A_{22}B_{22}$ [5]

Interphase behaviour for PVA bionanocomposites reinforced with HNTs and Cloisite 30B clays is similar to that in PVA/NBC bionanocomposites. Besides, $t_{\text{Interphase}}$ is independent of HNT diameter (D_{HNT}) and Cloisite 30B clay thickness ($t_{\text{Cloisite 30B}}$), as depicted in Fig. 5.9a, b accordingly. Furthermore, $L_{\text{Interphase-max}}$ in both bionanocomposite systems is also enlarged in a monotonic manner when increasing L_{HNT} and $L_{\text{Cloisite 30B}}$ in Fig. 5.9c, d, respectively. Similarly, with increasing DT_{HNT} and $W_{\text{Cloisite30B}}$, $W_{\text{Interphase-max}}$ is also enhanced in case of PVA/HNT bionanocomposites and PVA/Cloisite 30B clay bionanocomposites accordingly, as observed in Fig. 5.9e, f. Conversely, $H_{\text{Interphase-max}}$ reveals a linearly decreasing trend when DL_{HNT} and $H_{\text{Cloisite 30B}}$ become greater shown in Fig. 5.9g, h.

A proposed mechanism model in terms of effects of nanofiller dispersion and particle–matrix interaction is demonstrated in Fig. 5.1. It is identified in this study that two typical categories for the particle–matrix interaction can be addressed including fully embedded and partially embedded nanoparticles within PVA matrices. The former yields more effective interphase surrounding entire nanoparticles (i.e. full

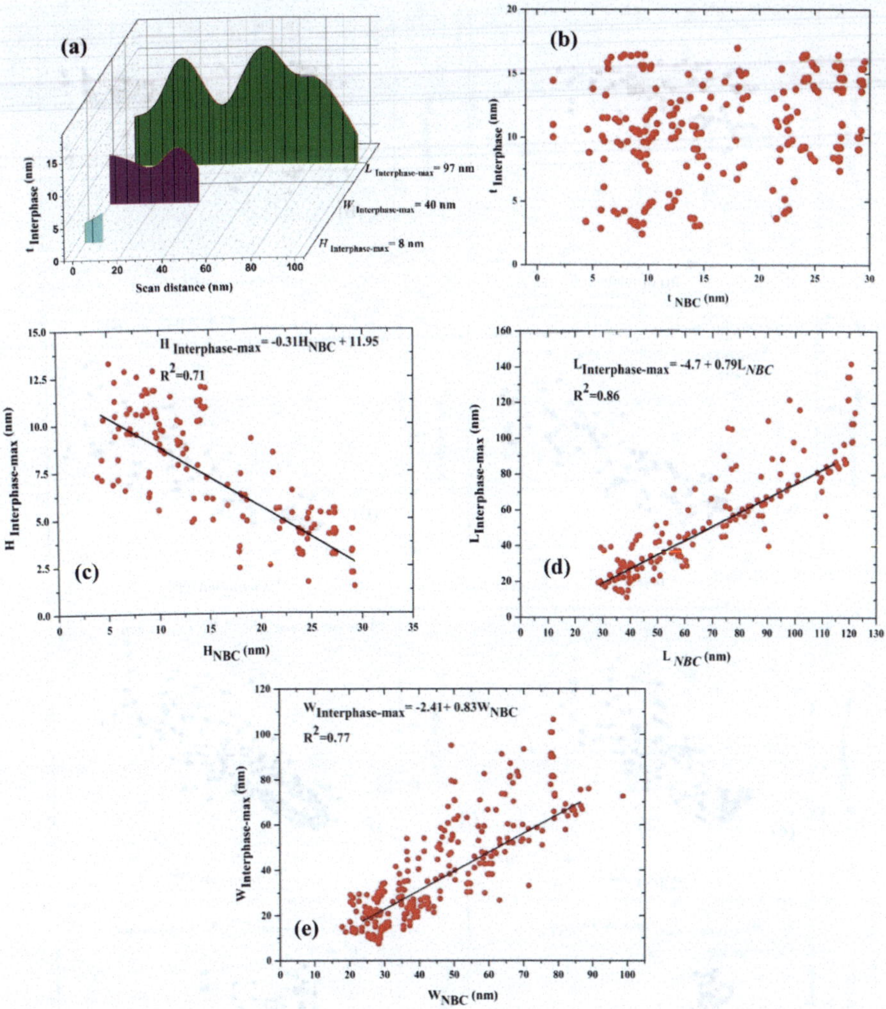

Fig. 5.8 **a** Maximum interphase dimensions along typical PVA/NBC interphase and relationships between NBC and interphase dimensions in PVA/NBC bionanocomposites: **b** interphase thickness ($t_{\text{Interphase}}$) and NBC thickness (t_{NBC}), **c** maximum interphase height ($H_{\text{Interphase-max}}$) and NBC height ($H_{\text{NBC}}$), **d** maximum interphase length ($L_{\text{Interphase-max}}$) and NBC length ($L_{\text{NBC}}$), as well as **e** maximum interphase width ($W_{\text{Interphase-max}}$) and NBC width ($W_{\text{NBC}}$) [3]

coverage of nanoparticles by the interphases), which is considered as a good representation for effective load transfer from matrices to nanofillers. The latter offers less pronounced interphase effectiveness since some remarkable portions of nanoparticles are free of contact with PVA matrices, leading to the reduction of effective interfacial regions. Interphase volume $V_{\text{Interphase}}$ is regarded as an essential characteristic

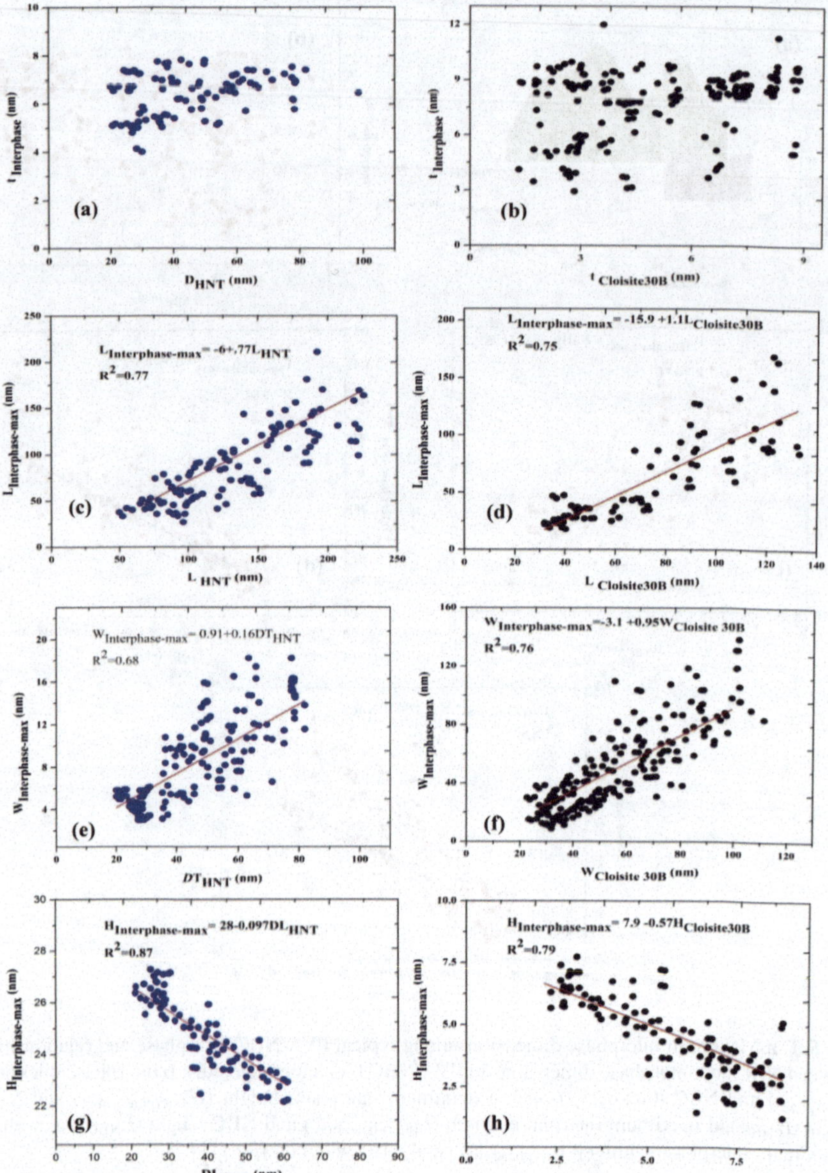

Fig. 5.9 Relationship between interphase dimensions and nanofiller dimensions: **a**, **b** interphase thickness ($t_{\text{Interphase}}$) with **a** HNT diameters (D_{HNT}) and **b** Cloisite 30B clay thickness ($t_{\text{Cloisite 30B}}$), **c**, **d** maximum interphase length ($L_{\text{Interphase-max}}$) with **c** HNT length (L_{HNT}) and **d** Cloisite 30B clay length ($L_{\text{Cloisite 30B}}$), **e**, **f** maximum interphase width ($W_{\text{Interphase-max,}}$) with **e** HNT transverse diameters (DT_{HNT}) and **f** Cloisite 30B clay width ($W_{\text{Cloisite 30B}}$) and **e**, **f** maximum interphase height ($H_{\text{Interphase-max}}$) with **g** HNT longitudinal diameters (DL_{HNT}) and **h** Cloisite 30B clay height ($H_{\text{Cloisite 30B}}$) [5]

parameter to evaluate the effects of fully or partially embedded nanoparticles in PVA-based bionanocomposites, which can be specified as $(V_{\mathrm{Interphase}})_f$ and $(V_{\mathrm{Interphase}})_p$, respectively. Figures 5.10c and 5.11a, b display the interphase modulus ($E_{\mathrm{Interphase}}$)

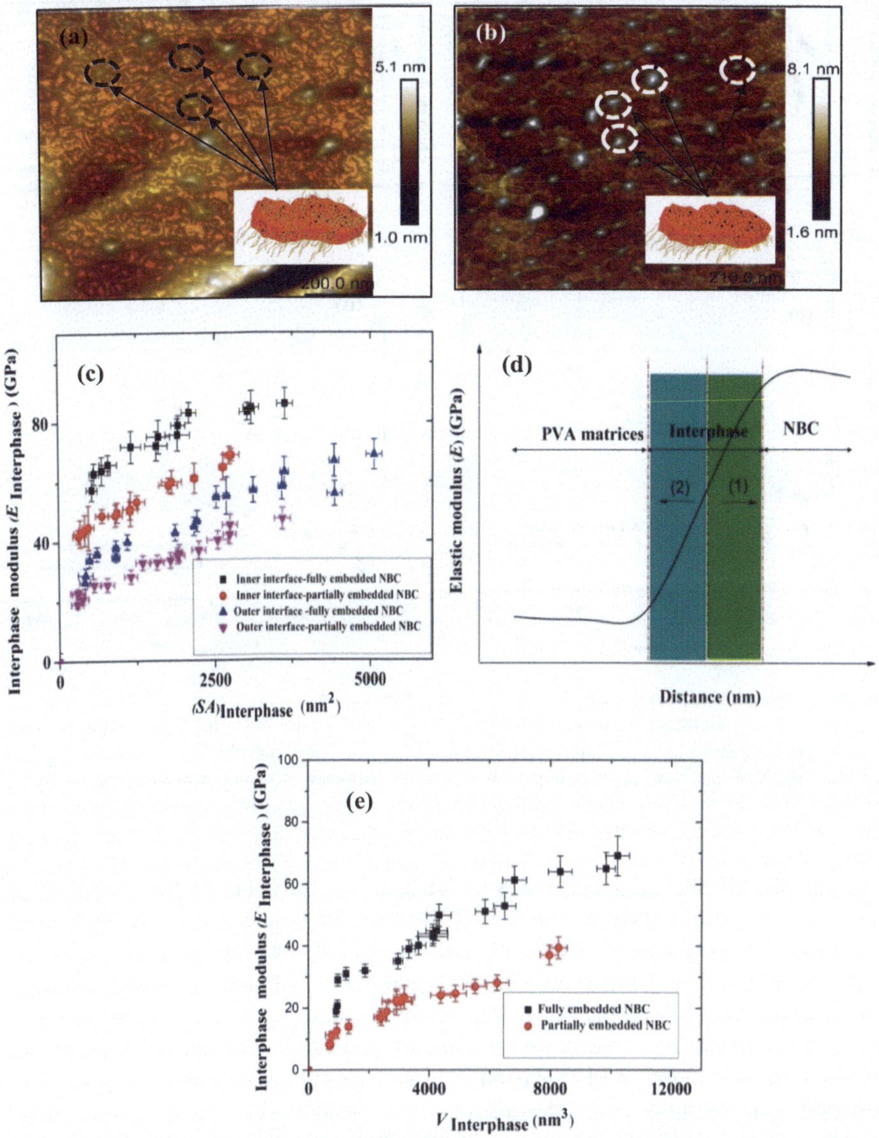

Fig. 5.10 2D AFM height mapping images of PVA/NBC bionanocomposites: **a** fully embedded NBCs and **b** partially embedded NBCs, **c** relationship between interphase area ($SA_{\mathrm{Interphase}}$) and interphase modulus ($E_{\mathrm{Interphase}}$), **d** schematic diagram for two proposed interphase zones and **e** relationship between interphase volume ($V_{\mathrm{Interphase}}$) and interphase modulus ($E_{\mathrm{Interphase}}$) [3]

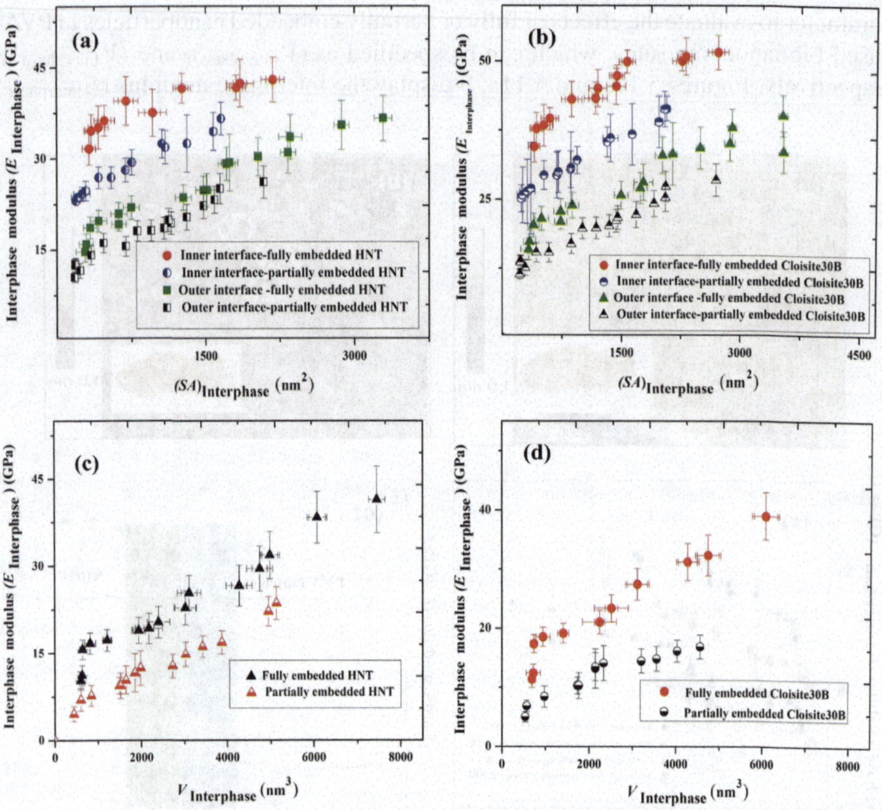

Fig. 5.11 Relationships between interphase surface area ($SA_{Interphase}$) and interphase modulus ($E_{Interphase}$) in **a** PVA/HNT bionanocomposites and **b** PVA/Cloisite 30B clay bionanocomposites, as well as relationships between interphase volume ($V_{Interphase}$) and interphase modulus ($E_{Interphase}$) in **c** PVA/HNT bionanocomposites and **d** PVA/Cloisite 30B clay bionanocomposites, respectively [5]

as a function of $SA_{outer\ interface}$ and $SA_{inner\ interface}$ for fully and partially embedded nanoparticles within PVA matrices in PVA/NBC bionanocomposites, PVA/HNT bionanocomposites and PVA/Cloisite 30B clay bionanocomposites, respectively. All three bionanocomposite types demonstrated the same behaviour with increasing $SA_{outer\ interface}$ and $SA_{inner\ interface}$. Additionally, $E_{Interphase}$ can be improved remarkably in a nonlinear manner accordingly. In particular, fully embedded NBCs, HNTs and Cloisite 30B clays yield consistently higher interphase modulus as opposed to corresponding partially embedded nanoparticles when either inner or outer interface is considered. Elastic properties of interphase are strongly affected by interphase dimensions since interphase modulus tends to become higher with increasing the surface areas of outer and inner interfaces Furthermore, as for fully embedded NBCs, HNTs and Cloisite 30B clays, a higher modulus-increasing trend was evident as opposed to that based on those partially embedded nanoparticles. This finding suggests that interphases can play a more important role in enhancing mechanical properties of PVA nanocomposites with better bonded matrix–filler interactions. Moreover, it has also been demonstrated that the surface area of inner interface gives rise to relatively high overall interphase modulus in contrast with corresponding surface area of outer interface. As mentioned earlier by Liu et al. [13], the interphase zone with modulus-gradient effect can be divided into two different regions, namely Region 1 and Region 2 for a typical case in Fig. 5.10d, in which Region 1 is in contact with nanoparticle zones with relatively high interphase density when compared with Region 2. Additionally, Fan et al. [17] reported that increasing the interphase density inevitably led to the modulus enhancement for the same type of materials, which was in good accordance with high interphase modulus results obtained in case of the surface area of inner interface. Apparently, the right boundary of Region 1 where inner interface is located is adjacent to nanoparticles zones with higher density and elastic modulus. The relationship between interphase modulus and interphase volume is demonstrated in Fig. 5.10e, Fig. 5.11c, d in case of PVA/NBC bionanocomposites, PVA/HNT bionanocomposites and PVA/Cloisite 30B clay bionanocomposites, respectively, on the basis of fully and partially embedded nanoparticles in PVA-based bionanocomposites. It is clearly identified that the significant enhancement of interphase modulus takes place with increasing interphase volume, which is similar to the case by increasing $SA_{Interphase}$. The same applies to fully embedded nanoparticles, resulting in much greater interphase modulus than partially embedded counterparts in terms of interphase volume in PVA-based bionanocomposites.

Aspect ratio is used as an important factor to evaluate the reinforcement efficiency of nanofillers within polymer matrices. The determination of aspect ratio can be much easier when dealing with nanofillers in regular size and shape. However, in reality, irregular nanofillers in different shapes often occur leading to the complexity of aspect ratios due to the effect of multi-stage dispersion processing techniques used in manufacturing nanocomposites [1]. In addition to aspect ratios of nanofillers, interphase dimensions and properties can also be affected by the nanofiller dispersion. Liu and Brinson [1] and Alishahi et al. [18] suggested based on theoretical modelling that an effective parameter in controlling the reinforcement efficiency of nanofillers is the interphase volume per unit nanofiller volume ($V_{\text{interphase}}/V_{\text{nanofillers}}$) instead of $V_{\text{interphase}}$ alone. In this work, $V_{\text{interphase}}/V_{\text{nanofillers}}$ has been utilised in PVA bionanocomposites for all three nanofillers used with corresponding results being presented in Fig. 5.12. It is clearly observed that with increasing $V_{\text{nanofillers}}$, $V_{\text{interphase}}/V_{\text{nanofillers}}$ significantly decreases for PVA bionanocomposites reinforced with NBCs and Cloisite 30B clays though this tendency becomes less pronounced for PVA/HNT bionanocomposites. Overall, the inclusion of NBCs in PVA bionanocomposites induces the highest values of $V_{\text{interphase}}/V_{\text{nanofillers}}$ when compared with those of HNTs and Cloisite 30B clays. This finding suggests the most effective reinforcement efficiency of NBCs leading to the highest mechanical properties of PVA/NBC bionanocomposites.

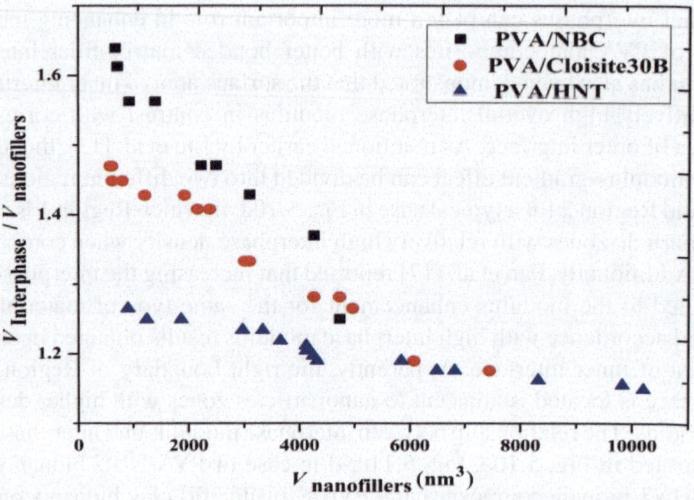

Fig. 5.12 $V_{\text{interphase}}/V_{\text{nanofillers}}$ as a function of $V_{\text{nanofillers}}$ for PVA/NBC bionanocomposites, PVA/Cloisite 30B clay bionanocomposites and PVA/HNT bionanocomposites

5.2.4 *Particle Debonding*

When NBC loading increases up to 5 wt%, interparticle spacing appears to be smaller, and NBCs tend to be interconnected with neighbouring particles side by side, illustrated in Fig. 5.13a, in which contact-free particles are clearly seen on some sides within PVA matrices. Furthermore, a joint particle-to-particle formation may induce the phase separation leading to scattered load transfer from PVA matrices to NBCs, as evidenced by the particle overlap of NBCs in Fig. 5.13b. At the NBC loading of 10 wt%, NBC aggregation becomes manifested with a clear sign of large stacked-up NBC agglomerates, displayed in Fig. 5.13c. As such, the particle-to-particle separation occurs more evidently with resulting weak interfacial bonding. Figure 5.13d reveals the modulus-variation mapping on the surface cross section of a line scan (i.e. the line between two ovals $A_{23}B_{23}$ in Fig. 5.13c), which highlights two narrow zones with a sharp decline in elastic modulus. In addition to particle dimensions, particle dispersion and distribution patterns in polymer matrices are the other key factors in controlling interfacial bonding between NBCs and PVA matrices in PVA/NBC bionanocomposites. When excessive amounts of NBCs are dispersed within PVA matrices, NBCs tend to agglomerate owing to their weak van der Waals interactions. Additionally, increasing the particle loading inevitably decreases the interparticle spacing, and thus hinders good particle dispersion [19]. Blighe et al. [20] reported that the interparticle spacing was determined to be 6 nm in PVA nanocomposites reinforced with 10 vol% single-walled CNTs (SWNTs) leading to typical particle aggregation. As a result, a slippage phenomenon of stacked NBCs under mechanical loading gives rise to a less effective enhancement level for mechanical properties of nanocomposites. Moreover, particle agglomeration becomes apparent at

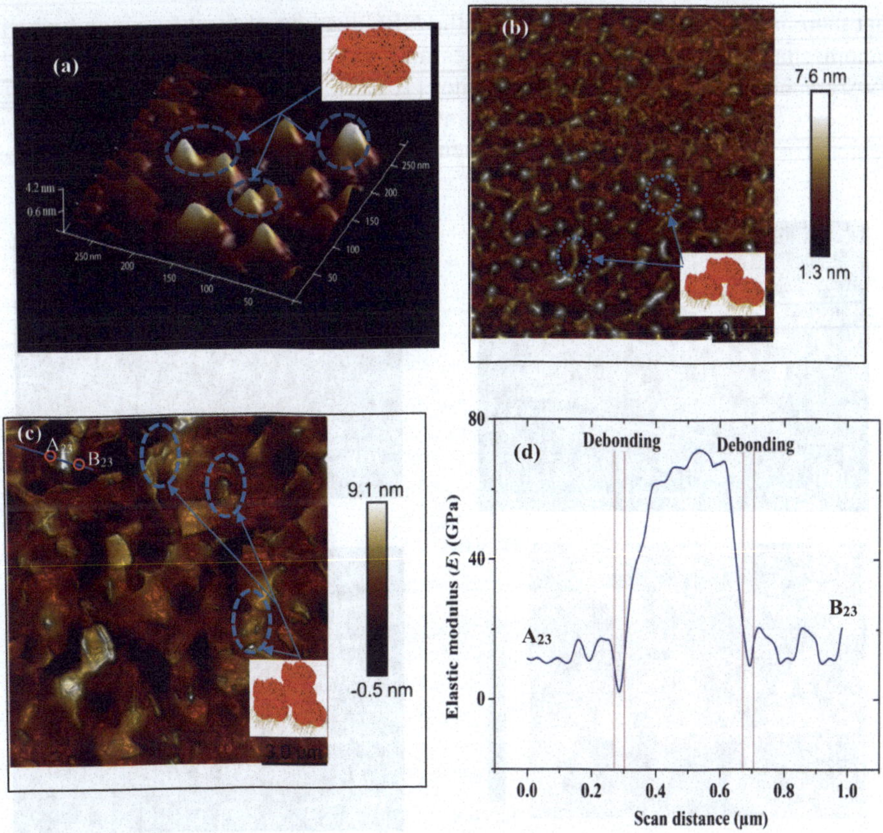

Fig. 5.13 **a** Typical 3D AFM height mapping image of PVA/5 wt% NBC bionanocomposites with different surface effects: **a** NBC joint edges and **b** NBC overlapping, **c** 2D height mapping image of PVA/10 wt% NBC bionanocomposites for phase separation and **d** modulus mapping profile of PVA/10 wt% NBC bionanocomposites cut along the cut section $A_{23}B_{23}$ [3]

high NBC contents depicted in Fig. 5.13. Such a finding can be explained by Li et al. [14] that in a nanocomposite system, nanoparticles are brushed with modified layers of polymer matrices especially at a high particle volume fraction. Additionally, 'modified polymer shells' surrounding different particles overlap to form continuous phases, which means that such modified polymer shells become the interphases with separated regions from 'parents' matrices', resulting in less desirable properties. Furthermore, some NBCs are unable to actively interact with molecular chains of PVA leading to weaker interfacial bonding and higher possibility of phase separation between NBCs and PVA matrices. This result is indicative of particle debonding effect taking place between nanofillers and polymer matrices [21]. It is also worth mentioning that lower interphase modulus can be associated with higher densities of nanoparticles and polymer matrices, as compared to that of interphase zones [4]. The aforementioned NBC dispersion pattern in PVA/NBC bionanocomposites suggests that there is a percolation threshold for the NBC loading of 5 wt%, beyond which nanomechanical properties of interphases may diminish in good accordance with PVA/graphene oxide (GO) nanocomposites [10, 22].

In case of PVA bionanocomposites reinforced with Cloisite 30B clays and HNTs depicted in Fig. 5.14a, d, a very similar phenomenon was also noticed that the distance

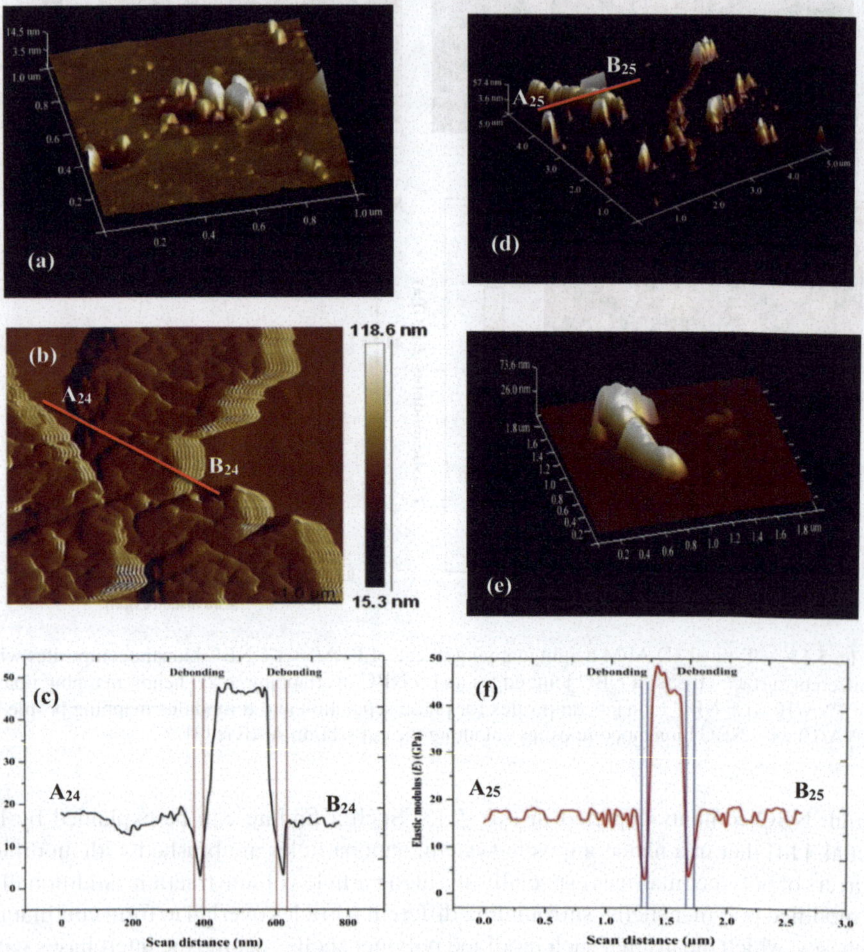

Fig. 5.14 **a** Typical 3D height mapping image of PVA/5 wt% Cloisite 30B clay bionanocomposite surfaces, **b** 2D height mapping image of PVA/5 wt% Cloisite 30B clay bionanocomposite surfaces for phase separation [5], **c** elastic modulus mapping profile of PVA/10 wt% Cloisite 30B clay bionanocomposite sample cut along the cut section $A_{24}B_{24}$ [5], **d** 3D height mapping image of PVA/5 wt% HNT bionanocomposite surfaces for phase separation, **e** 3D height mapping image of PVA/10 wt% HNT bionanocomposite surfaces for HNT agglomeration and **f** elastic modulus mapping profile of PVA/10 wt% HNT bionanocomposite sample along the cut section $A_{25}B_{25}$ in **d** [5]

between nanofillers of both Cloisite 30B clays and HNTs decreases to cause particle overlap and agglomeration when nanofiller loading increases accordingly. It is very clearly shown in Fig. 5.14b that Cloisite 30B clay aggregates reveal typical stacks of layered structures. On the other hand, the debonding effect takes place with the addition of excessive amounts of Cloisite 30B clays and HNTs, Fig. 5.14c, f, respectively. Nonetheless, as seen from Fig. 5.14b, d, e, Cloisite 30B clays in large stacks of layered structures, as well as HNTs seem to easily form large particle agglomerates as opposed to NBCs. Such results can be associated with their relatively high aspect ratios prone to considerable particle waviness to deteriorate their reinforcement efficiency, thus resulting in less favourable mechanical properties of corresponding PVA nanocomposites. In addition, nanofiller dispersion methods such as ultrasonication [18] can also give rise to possible structure damage of nanofillers with high ultrasonic power intensity resulting in detrimental properties, as evidenced by AFM images of Cloisite 30B clays and HNTs in corresponding Fig. 5.3a and 5.4a.

5.3 Summary

In this chapter, the following key points can be summarised:

- A pioneering approach has been developed for determining nanoelastic behaviour of PVA nanocomposites, as well as 3D interphase dimensions and interphase moduli in terms of surface area and interphase volume.
- Experimental characterisation results indicate that interphase thickness appears to be nonuniform among individual particles, but becomes independent of particle thickness.
- Interphase modulus in PVA-based nanocomposite system is enhanced with increasing interphase volume. Moreover, interphase modulus in case of PVA/NBC nanocomposites is higher than those for PVA/Cloisite 30B clay nanocomposites and PVA/HNT nanocomposites with respect to $V_{\text{interphase}}/V_{\text{nanofillers}}$.
- Nanomechanical properties and sizes of interphases depend greatly on nanofiller dispersion state. Nevertheless, increasing nanofiller content has been found to result in the decrease in intraparticle spacing leading to particle agglomeration with the interfacial debonding between PVA matrices and nanoparticles.
- Two key geometric parameters comprising interphase surface area and interphase volume have been detected to be non-constant variables, which can be potentially influenced by nanofiller dimensions and sizes, as well as nanofiller dispersion states. Our approach enlightens a more effective approach to consider actual interphase properties in theoretical and numerical modelling framework instead of using a simple assumption of single interphase layer dimensions and uniform interphase properties in order to achieve more accurate prediction for overall properties of nanocomposite systems.

References

1. Liu H, Brinson LC (2008) Reinforcing efficiency of nanoparticles: a simple comparison for polymer nanocomposites. Compos Sci Technol 68(6):1502–15012
2. Mousa M, Dong Y (2017) Strong poly (vinyl alcohol)(PVA)/bamboo charcoal (BC) nanocomposite films with particle size effect. ACS Sustain Chem Eng 6(1):467–479
3. Mousa M, Dong Y (2018) Novel three-dimensional interphase characterisation of polymer nanocomposites using nanoscaled topography. Nanotechnology 29(38):385701
4. Behmer DJ, Hawkins CP (1986) Effects of overhead canopy on macroinvertebrate production in a Utah stream. Freshw Biol 16(3):287–300
5. Mousa M, Dong Y (2020) Towards sophisticated 3D interphase modelling of advanced bionanocomposites via atomic force microscopy. J Nanomater Article ID 4526108
6. Kim J-K, Mai Y-W (1998) Engineered interfaces in fiber reinforced composites. Elsevier, Kidlington
7. Tashiro K, Tadokoro H (1981) Calculation of three-dimensional elastic constants of polymer crystals. 3. α and γ forms of nylon 6. Macromolecules 14(3):781–785
8. Tashiro K, Kobayashi M (1991) Theoretical evaluation of three-dimensional elastic constants of native and regenerated celluloses: role of hydrogen bonds. Polymer 32(8):1516–1526
9. Mishra SK, Kannan S (2014) Development, mechanical evaluation and surface characteristics of chitosan/polyvinyl alcohol based polymer composite coatings on titanium metal. J Mech Behav Biomed Mater 40:314–324
10. Jafarzadeh S, Claesson PM, Sundell PE, Pan J, Thormann E (2014) Nanoscale electrical and mechanical characteristics of conductive polyaniline network in polymer composite films. ACS Appl Mater Interfaces 6(21):19168–19175
11. Clifford CA, Seah MP (2005) Quantification issues in the identification of nanoscale regions of homopolymers using modulus measurement via AFM nanoindentation. Appl Surf Sci 252(5):1915–1933
12. Syed Asif S, Wahl KJ, Colton RJ, Warren OL (2001) Quantitative imaging of nanoscale mechanical properties using hybrid nanoindentation and force modulation. J Appl Phys 90(3):1192–1200
13. Liu Y, Hamon AL, Haghi-Ashtiani P, Reiss T, Fan B, He D, Bai J (2016) Quantitative study of interface/interphase in epoxy/graphene-based nanocomposites by combining STEM and EELS. ACS Appl Mater Interfaces 8(49):34151–34158
14. Li Y, Huang Y, Krentz T, Natarajan B, Neely T, Schadler LS (2016) Polymer nanocomposite interfaces: the hidden lever for optimizing performance in spherical nanofilled polymers. In: Netravali NA, Mittal KL (eds) Interface/interphase in polymer nanocomposites. Scrivener Publishing LLC, Beverly, pp 1–69
15. Widjojo N, Chung TS, Weber M, Maletzko C, Warzelhan V (2011) The role of sulphonated polymer and macrovoid-free structure in the support layer for thin-film composite (TFC) forward osmosis (FO) membranes. J Membr Sci 383(1–2):214–223
16. Mousa MH, Dong Y, Davies IJ (2016) Recent advances in bionanocomposites: preparation, properties, and applications. Int J Polym Mater Polym Biomater 65(5):225–254
17. Fan H, Hartshorn C, Buchheit T, Tallant D, Assink R, Simpson R, Kissel DJ, Lacks DJ, Torquato S, Brinker CJ (2007) Modulus–density scaling behaviour and framework architecture of nanoporous self-assembled silicas. Nat Mater 6(6):418–423
18. Alishahi E, Shadlou S, Doagou-R S, Ayatollahi MR (2013) Effects of carbon nanoreinforcements of different shapes on the mechanical properties of epoxy-based nanocomposites. Macromol Mater Eng 298(6):670–678
19. Jesson DA, Watts JF (2012) The interface and interphase in polymer matrix composites: effect on mechanical properties and methods for identification. Polym Rev 52(3):321–354
20. Blighe FM, Young K, Vilatela JJ, Windle AH, Kinloch IA, Deng L, Young RJ, Coleman JN (2011) The effect of nanotube content and orientation on the mechanical properties of polymer–nanotube composite fibers: separating intrinsic reinforcement from orientational effects. Adv Funct Mater 21(2):364–371

21. Gu Y, Li M, Wang J, Zhang Z (2010) Characterization of the interphase in carbon fiber/polymer composites using a nanoscale dynamic mechanical imaging technique. Carbon 48(11):3229–3235
22. Zhao X, Zhang Q, Chen D, Lu P (2010) Enhanced mechanical properties of graphene-based poly (vinyl alcohol) composites. Macromolecule 43(5):2357–2363

Chapter 6
Micromechanical Models of PVA-Based Bionanocomposite Films

Abstract To investigate the effect of the interphase on elastic modulus of bulk bionanocomposite films, a theoretical framework was proposed for the first time to systematically calculate interphase volume fraction according to stereological theory and nearest-surface distribution functions in this chapter. A three-phase composite model was employed based on hard-core-soft-shell structures including hard monodispersed or polydispersed anisotropic particles, as well as soft interphase and matrices. The effective volume fraction of nanoparticles was estimated in a simple theoretical approach by considering both volume fraction and geometric configuration of interphases. Experimentally determined tensile moduli of polyvinyl alcohol (PVA)-based bionanocomposite films were predicted by using Halpin–Tsai model and Mori–Tanaka model in which effective volume fraction of randomly oriented nanoparticles resulted from the inclusion of interphase properties and volume fractions. Moreover, associated results suggested that the estimation of elastic modulus according to effective volume fraction revealed much better agreement with experimental data, as opposed to that based on nominal volume fraction. In particular, the use of polydispersed nanodiameter bamboo charcoals (NBCs), halloysite nanotubes (HNTs) and Cloisite 30B clays with Fuller particulate gradation was proven to show the best prediction with experimental data among all proposed theoretical models.

Keywords Micromechanical modelling · Nanomechanical properties · 3D interphase · Monodispersed and polydispersed particles · Particulate gradation

6.1 Introduction

Micromechanical models such as Halpin–Tsai model [1] and Mori–Tanaka model [2] generally neglect the interphase formation between fillers and matrices for simplicity. As a result, mechanical properties of polymer nanocomposites can be either overestimated based on such conventional composite theoretical models for polymer/carbon

---

This chapter takes partial content materials from authors' published research article **'Mousa M, Dong Y (2020) Towards sophisticated 3D interphase modelling of advanced bionanocomposites via atomic force microscopy, J Nanomater, Article ID 4526108'** under Creative Commons Attribution License.

M. Mousa and Y. Dong, *Multiscaled PVA Bionanocomposite Films*,
https://doi.org/10.1007/978-981-15-8771-9_6

nanotube (CNT) composites [3] and silk fibroin/graphene oxide (GO) nanocomposites [4], or underpredicted for polyamide 66 (PA66)/calcium carbonate ($CaCO_3$) nanocomposites [5]. Such drawbacks in modelling work for polymer nanocomposites are inevitably associated with the absence of interphase features and properties. Many recent attempts have been made by incorporating interphase regions as the third phase within available micromechanical models like Mori–Tanaka model [6] and Maxwell model [7] with little success. This can be attributed to simple assumption that the interphase can be one-dimensional transitional material phase surrounding nanofillers with uniform interphase thickness and regular shapes while in reality three-dimensional (3D) irregular-shape interphase with non-uniform thickness generally occurs in the material morphology of nanocomposites [8].

In order to consider the benefit of real interphase effect in nanocomposite modelling work, interphase volume fraction is one of the key interphase material parameters to be predetermined. For instance, with the addition of 5 vol% monodispersed spherical nanoparticles (particle diameter of 10 nm and interphase thickness of 0.5 nm), the volume fraction of interphase can be as high as 25 vol% [9]. More impressively, when particle diameter is reduced to less than 5 nm, the volume fraction of interphase is increased by over 50 vol%, as compared to that of particles [9]. However, a full understanding for the effect of interphase volume fraction on elastic modulus of bulk nanocomposites still remains challenging owing to complex interphase network structures and an overlapping potential to neighbouring interfacial layers. The main objective of this chapter is to investigate effect of different nanofiller shapes and structures on nanomechanical properties of 3D interphase for finally prepared PVA-based bionanocomposites. Moreover, mechanical properties of PVA-based bionanocomposite were predicted by considering experimentally measured dimensions and volume fraction of interphase. In particular, polyvinyl alcohol (PVA) was selected as typical biopolymer after a great environmental concern associated with using conventional non-degradable petroleum-based polymers. Additionally, three different types of nanoparticles including Cloisite 30B clays, halloysite nanotubes (HNTs) and nanodiameter bamboo charcoals (NBCs) were chosen as typical nanofillers in different shapes and sizes.

6.2 Theory

6.2.1 *Micromechanical Models Based on Nominal and Effective Volume Fractions*

In general, the prediction of elastic modulus in case of unidirectional or randomly distributed filler reinforced composites can be investigated by means of conventional composite theoretical models such as Halpin–Tsai model. Such a model could also be employed to estimate elastic moduli of PVA-based bionanocomposites with different nanofiller orientation states, namely well-aligned and randomly oriented nanofillers, which are denoted as E_{random} and E_{parallel}, respectively, in the following equations:

$$E_{random} = E_m\left[\frac{3}{8}\left(\frac{1+\eta_L\xi\varnothing_p}{1-\eta_L\varnothing_p}\right)+\frac{5}{8}\left(\frac{1+2\eta_T\varnothing_p}{1-\eta_T\varnothing_p}\right)\right] \tag{6.1}$$

$$E_{\text{parallel}} = E_{\text{m}}\left[\frac{1+\eta_{\text{L}}\xi\varnothing_{\text{p}}}{1-\eta_{\text{L}}\varnothing_{\text{p}}}\right] \tag{6.2}$$

$$\eta_{\text{L}} = \frac{(E_{\text{p}}/E_{\text{m}})-1}{(E_{\text{p}}/E_{\text{m}})+\xi} \tag{6.3}$$

$$\eta_{\text{T}} = \frac{(E_{\text{p}}/E_{\text{m}})-1}{(E_p/E_{\text{m}})+2} \tag{6.4}$$

$$\xi = \frac{2\alpha}{3} = \frac{2l_{\text{p}}}{3t_{\text{p}}} \tag{6.5}$$

$$\varnothing_{\text{p}} = \frac{W_{\text{p}}}{W_{\text{p}}+(\rho_{\text{p}}/\rho_{\text{m}})(1-W_{\text{p}})} \tag{6.6}$$

where E_{random} and E_{parallel} represent Young's moduli of PVA-based bionanocomposites reinforced with randomly oriented nanofillers and well-aligned nanofillers parallel to the material surface under unidirectional loading, respectively. E_{m} and E_{p} are Young's moduli of PVA matrices and nanoparticles (i.e. NBCs, HNTs and Cloisite 30B clays) accordingly. α, l_{p} and t_{p} refer to the aspect ratio, the length and the thickness of nanoparticles. ϕ_{p} and W_{p} are nominal volume fraction and weight fraction of nanofillers in PVA-based bionanocomposites, respectively. ρ_{p} and ρ_{m} are the densities of nanoparticles and PVA accordingly. Furthermore, Mori–Tanaka model is also an effective theoretical model to predict elastic moduli of polymer nanocomposites. It was originally employed by Tandon and Weng [10] using Mori–Tanaka theory [2] and Eshelby's solution [11] to derive a complete analytical solution for calculating elastic moduli of an isotropic matrix containing aligned spheroidal inclusions. Longitudinal and transverse elastic moduli of polymer nanocomposites (E_{11}) and (E_{22}) can be expressed as follows:

$$\frac{\text{E}_{11}}{\text{E}_m} = \frac{1}{1+\varnothing_p(\text{A}_1+2\nu_m\text{A}_2)/\text{A}} \tag{6.7}$$

$$\frac{E_{22}}{E_{\text{m}}} = \frac{1}{1+\varnothing_{\text{p}}(-2\nu_{\text{m}}A_3+(1-\nu_{\text{m}})A_4+(1+\nu_{\text{m}})A_5A)/2A} \tag{6.8}$$

where ν_{m} is the Poisson's ratio of polymer matrices and A_i (i = 1–5) represents the functions of Eshelby's tensors and properties of the matrices and fillers such as Young's modulus, Poisson's ratio, aspect ratio and volume fraction of fillers. When nanofillers are inclined to more random orientation, elastic modulus of nanocomposites can be further predicted by using the combination of laminate theory [12] and

Mori–Tanaka model given by:

$$E_c \cong 0.375E_{11} + 0.625E_{22} \tag{6.9}$$

In a nanocomposite system, large specific surface area of nanofillers renders the amounts of polymeric chains to attach to nanofiller surfaces in order to generate the interphase in nanocomposites, which means that nominal volume fraction of nanofillers may be altered when reinforced within polymer matrices. As such, effective volume fraction should be used instead for more accurate property prediction in modelling work. Wan and Chen [13] and other co-workers [11, 14] suggested the following equation to calculate effective volume fraction of nanofillers:

$$\emptyset'_p = \emptyset_p(1 + kxR_g A_T \rho_p) \tag{6.10}$$

where A_T is the specific surface area. kx is used for the layer of absorbed polymeric chains in nanoparticles with a thickness xR_g of the polymer, which means the layer thickness is increased to x times of the radius of gyration of polymer R_g. The value of kx can be changed for different nanocomposite systems due to the variation of interfacial interactions between polymer matrices and nanofillers. Besides, the dynamic behaviour of polymeric chains near attractive interfaces would be different from those bulk polymeric phases and the effect of attractive surfaces could be held for the distances larger than the radius of gyration for polymers [15].

The term of kx can be reconsidered based on the equation [11, 13] given below:

$$kx = \frac{E_m}{E_p}\left(\frac{t}{Rg}\right) \tag{6.11}$$

where t is regarded as the thickness of adsorbed layers.

In polymer/platelet-like filler nanocomposites, $\emptyset'_p$ can depend on intercalation and exfoliation levels of platelet-like fillers taking place within polymer matrices. With respect to clay intercalated structures in nanocomposites, $\emptyset'_p$ can be calculated using the derived equation [10] shown as follows:

$$\frac{1}{\emptyset'_p} = 1 + \frac{\rho'_p(1 - W'_p W''_p S)[d'_1(N-1) + h]}{W'_p \rho_p [d_2(N-1) + h]} \tag{6.12}$$

where ρ'_p is referred to as the density of organoclays. W'_p is the weight fraction of organoclays in nanocomposites, and W''_p is the weight fraction of clay platelets within organoclays. S is the polymer uptake in the generic expression $S = (1 - W^1_p)/W^1_p$ to represent the mass of polymers swollen into the unit mass of intercalated clays where W^1_p is the weight fraction of intercalated clay structures. d'_1 is the basal plane spacing of organoclays. N and h are the average layer number per stack and the thickness of platelets, respectively. In case of exfoliated clay structures, when clay platelets are dispersed homogeneously in polymer matrices, the effective volume fraction can be calculated according to Eq. (6.10). The main difference from conventional nanocomposites is the adsorption effect. In Eq. (6.10), A_{T} becomes total surface area (A_{Tp}) of clay platelets, and ρ_{p} is converted into the density of clay platelets (ρ_{p}''). Hence, Eq. (6.10) can be rewritten below:

$$\text{Ø}'_p = \text{Ø}_p\left(1 + kxR_g \mathrm{A_{Tp}}\rho''_p\right) \tag{6.13}$$

6.2.2 *Micromechanical Models Based on Volume Fractions of Nanofillers and Interphase*

The interphase is very critical in nanocomposite systems, which cannot be completely ignored, as mentioned earlier in Chap. 5 since the interphase in PVA-based bionanocomposites has non-uniform thickness with their corresponding nanomechanical properties, which primarily depends on 3D interphase dimensions and features. For instance, when the interphase volume in PVA/NBC bionanocomposites increased from 2988 to 4363 nm^3, their interphase modulus was significantly enhanced by 44% from 34.67 to 49.91 GPa [9]. Consequently, the use of 3D interphase features in terms of interphase size and volume fraction to predict elastic modulus of polymer nanocomposites can lead to a more reliable modelling approach to estimate bulk nanocomposite properties according to their nanomechanical behaviour. A simple equation proposed to calculate the volume fraction of nanofillers based on interphase features is given below:

$$\text{Ø}_p = \frac{\text{Ø}_{\mathrm{Interphase}} V_p}{V_{\mathrm{Interphase}}} \tag{6.14}$$

where $\text{Ø}_{\mathrm{Interphase}}$ is the volume fraction of interphase surrounding anisotropic nanoparticles, which can be estimated by using the theory of nearest-surface distribution function [16] detailed in the next section. $V_{\mathrm{Interphase}}$ and V_{p} are the volumes of interphase and nanoparticles, respectively, which can be determined according to corresponding Eqs. (6.5) and (6.7).

6.2.3 *Interphase Volume Fraction ($\text{Ø}_{Interphase}$)*

Theoretical interfacial volume fraction can be derived from the nearest-surface distribution theory [16], in which composite media are composed of 3D rigid particles and voids. Lu and Torquato [16] employed statistical geometries of composites and geometric probability to determine a void exclusion probability $e_{\mathrm{v}}(r)$, which is defined as the probability of a designated empty region of composite media with the radius of particle cavities r at an arbitrary point. Accordingly, $e_v(r)$ can be given by

$$e_v(r) = (1 - \eta)\exp\left[-\pi\rho\left(er + dr^3 + gr^3\right)\right] \quad (6.15)$$

$$e = \frac{4\langle R^2\rangle}{1 - \eta} \quad (6.16)$$

$$d = \frac{4\langle R\rangle}{1 - \eta} + \frac{8\pi\rho\langle R^2\rangle^2}{(1 - \eta)^2} \quad (6.17)$$

$$g = \frac{4}{3(1 - \eta)} + \frac{16\pi\rho\langle R\rangle\langle R^2\rangle}{3(1 - \eta)^2} + \frac{64m\pi^2\rho^2\langle R^2\rangle^3}{27(1 - \eta)^3} \quad (6.18)$$

In practice, $e_v(r)$ represents the expected void fraction in such a two-phase composite system [17]. m is a parameter associated with the theoretical estimation of a radial distribution function in a spherical particle system. In particular, $m = 2$ has been used in this study based on Carnahan–Starling approximation [18]. ρ is the number density of isotropic particles with a radius R. η can be further calculated according to Eq. (6.19) as follows:

$$\eta = \rho\langle \mathrm{V}\rangle = \rho\frac{\pi^{3/2}R^3}{\Gamma(1 + 3/2)} = \frac{4\pi}{3}\rho\left\langle R^3\right\rangle \quad (6.19)$$

where $\langle V\rangle$ is the average volume of hard particles, $\Gamma(\mathrm{x})$ is the gamma function. According to quantitative stereology [19], η denotes the volume fraction of 3D rigid particles.

Composite materials are supposed to consist of a packing of rigid particles and matrices, $e_v(r)$, as mentioned earlier, can thus be considered as the matrices in a two-phase composite material. When a three-phase composite material, consisting of rigid nanoparticles, matrices and soft-shell interphases (layer thickness $r = t$) surrounding particles, is taken into consideration, the combination of each of isotropic particles and associated soft-shell interphase can be regarded as a composite particle. Therefore, a three-phase composite system can be simplified as a two-phase equivalent system including rigid composite particles (i.e. original particles and corresponding interphase), as well as matrices. Nonetheless, for such a three-phase composite system, $e_v(t)$ is not applicable for Eqs. (6.15)–(6.18), but can be subjected to the geometric configuration of anisotropic particles instead. The incorporation of geometric details of particles yields the modified equations given by:

$$e_v(t) = (1 - \eta)\exp\left[-\pi\rho\left(et + dt^3 + gt^3\right)\right] \quad (6.20)$$

$$e = \frac{\langle \mathrm{S}\rangle}{\pi(1 - \eta)} \quad (6.21)$$

$$d = \frac{4\langle R\rangle}{1-\eta} + \frac{\rho\langle R^2\rangle}{2\pi(1-\eta)^2} \tag{6.22}$$

$$g = \frac{4}{3(1-\eta)} + \frac{4\rho\langle R\rangle\langle S\rangle}{3(1-\eta)^2} + \frac{m\rho^2\langle S\rangle^3}{27\pi(1-\eta)^3} \tag{6.23}$$

where $\langle S\rangle$ is the average area of rigid particles. According to Eqs. (6.20)–(6.23), it can be clearly shown that $e_v(t)$ depends primarily on R, ρ and surface area of isotropic particles S. Similarly, in an anisotropic particle system, $e_v(t)$ should be calculated based on the geometric details of such particles. Hence, the key step is how to determine these characteristic parameters of anisotropic particles. According to the previous studies [20, 21], an equivalent diameter (D_{eq}) is widely used as an alternative dimension of anisotropic irregular particles with complex geometries. The number density of anisotropic particles can be estimated by a quantitative stereological theory [19], which is equal to the ratio of the volume fraction of solid phase to the average volume of rigid particles, or alternatively the ratio of specific surface area of solid phase to the average surface area of rigid particles. In terms of D_{eq}, average volume $\langle V\rangle$ and average surface area $\langle S\rangle$ of anisotropic particles can be calculated as

$$\langle V\rangle = \frac{\pi}{6}\langle D_{eq}^3\rangle \tag{6.24}$$

$$\langle S\rangle = \frac{\pi}{S}\langle D_{eq}^2\rangle \tag{6.25}$$

The volume fraction of solid phase with various geometries has been found to be relatively similar to the volume fraction of rigid particles [22]. As a result, the number density (N_V) of anisotropic particles is given by the stereological theory shown below

$$N_V = \frac{V_p}{\langle V\rangle} = \frac{S_V}{\langle S\rangle} = \frac{6V_p}{\pi\langle D_{eq}^3\rangle} \tag{6.26}$$

where S_V is the specific surface area of rigid anisotropic particles. By substituting Eqs. (6.24) and (6.25) into Eqs. (6.20)–(6.23), $e_v(t)$ for composite materials reinforced with anisotropic particles can be expressed as

$$e_v(t) = (1-V_p)\exp\left[-\frac{6V_p}{\langle D_{eq}^3\rangle}(e't + d't^3 + g't^3)\right] \tag{6.27}$$

$$\emptyset_{\text{interphase}} = 1 - V_p - e_v(t) = (1-V_p)\left\{1 - \exp\left[-\frac{6V_p}{\langle D_{eq}^3\rangle}(e't + d't^3 + g't^3)\right]\right\} \tag{6.28}$$

In a PVA-based bionanocomposite system, V_p represents the nominal volume fraction of nanofillers ($\emptyset_{\text{p-nominal}}$). As such, Eq. (6.28) can be rewritten as:

$$\emptyset_{\text{Interphase}} = \left(1 - \emptyset_{p\text{-nominal}}\right)\left\{1 - \exp\left[-\frac{6\emptyset_{p\text{-nominal}}}{\langle D_{eq}^3\rangle}\left(e't + d't^3 + g't^3\right)\right]\right\} \quad (6.29)$$

$$e' = \frac{\left\langle D_{\text{eq}}^2\right\rangle}{S\left(1 - \emptyset_{\text{p-nominal}}\right)} \quad (6.30)$$

$$d' = \frac{2\left\langle D_{\text{eq}}\right\rangle}{1 - \emptyset_{\text{p-nominal}}} + \frac{3\emptyset_{\text{p-nominal}}\left\langle D_{\text{eq}}^2\right\rangle^2}{S^2\left(1 - \emptyset_{\text{p-nominal}}\right)^2\left\langle D_{\text{eq}}^3\right\rangle} \quad (6.31)$$

$$g' = \frac{4}{3\left(1 - \emptyset_{\text{p-nominal}}\right)} + \frac{4\emptyset_{\text{p-nominal}} D_{\text{eq}}\left\langle D_{\text{eq}}^2\right\rangle}{S\left(1 - \emptyset_{\text{p-nominal}}\right)^2\left\langle D_{\text{eq}}^3\right\rangle} + \frac{4m\emptyset_{\text{p-nominal}}\left\langle D_{\text{eq}}^2\right\rangle^3}{3S^3\left(1 - \emptyset_{\text{p-nominal}}\right)^3\left\langle D_{\text{eq}}^3\right\rangle^2} \quad (6.32)$$

$$S = \frac{(1 + 1.5\alpha)^{2/3}}{1 + \alpha} \quad (6.33)$$

In a typical nanocomposite system, nanoparticles can be dispersed uniformly with relatively similar sizes, which are known as monodispersed nanoparticles [23]. On the contrary, when nanoparticles with different sizes and diameters are dispersed randomly, they are called polydispersed nanoparticles [23].

In case of a monodispersed particle system, the parameters e', d' and g' are expressed as:

$$e' = \frac{D_{eq}^2}{S\left(1 - \emptyset_{p\text{-nominal}}\right)} \quad (6.34)$$

$$d' = \frac{2D_{\text{eq}}}{1 - \emptyset_{\text{p-nominal}}} + \left[2 + \frac{3\emptyset_{\text{p-nominal}}}{S^2\left(1 - \emptyset_{\text{p-nominal}}\right)}\right] \quad (6.35)$$

$$g' = \frac{4}{3\left(1 - \emptyset_{\text{p-nominal}}\right)} + \left[\frac{1}{3} + \frac{\emptyset_{\text{p-nominal}}}{S\left(1 - \emptyset_{\text{p-nominal}}\right)} + \frac{m\emptyset_{\text{p-nominal}}^2}{3S^3\left(1 - \emptyset_{\text{p-nominal}}\right)^2}\right] \quad (6.36)$$

By substituting Eqs. (6.34)–(6.36) into Eq. (6.29), and then letting $\lambda = \frac{t'}{D_{\text{eq}}}$ where λ and t' are the geometric size factor of anisotropic particles and the interfacial dimension, respectively, one can obtain

$$\emptyset_{\text{Interphase}} = \left(1 - \emptyset_{p\text{-nominal}}\right)\left\{1 - \exp\left[-\frac{6\emptyset_{p\text{-nominal}}\lambda^3}{1 - \emptyset_{p\text{-nominal}}}\left[\frac{1}{S\lambda^2} + \frac{1}{\lambda}\left[2 + \frac{3\emptyset_{p\text{-nominal}}}{S^2\left(1 - \emptyset_{p\text{-nominal}}\right)}\right] + \frac{4}{3} + \frac{4\emptyset_{p\text{-nominal}}}{S\left(1 - \emptyset_{p\text{-nominal}}\right)} + \frac{4m\emptyset^2_{p\text{-nominal}}}{3S^3\left(1 - \emptyset_{p\text{-nominal}}\right)^2}\right]\right]\right\} \tag{6.37}$$

On the contrary, with respect to a polydispersed particle system, it is necessary to evaluate the effect of particle size distribution (PSD) on both the volume fraction of interphase and reinforcement efficiency of nanocomposites. The gradation of irregular aggregates can be experimentally investigated using a sieve analyser or a laser particle analyser where the size of each irregular aggregate is evaluated relative to the corresponding size of spherical particles [24]. The PSD of spherical particles can be transformed into that of anisotropic particles when the equivalent diameter is defined according to the previous research work [20]. In this study, two particulate gradations, namely Fuller gradation [22] and equal volume fraction (EVF) gradation [20, 22], were employed with their specific equations given below:

$$F_V(D) = \frac{\sqrt{D} - \sqrt{D_{\min}}}{\sqrt{D_{\max}} - \sqrt{D_{\min}}} \tag{6.38}$$

$$F_V(D) = \frac{\text{In}\,D - \text{In}\,D_{\min}}{\text{In}\,D_{\max} - \text{In}\,D_{\min}} \tag{6.39}$$

where $F_V(D)$ is the cumulative volume percentage of isotopic particles with their corresponding diameter of D, in which $D_{\max}$ and $D_{\min}$ are the maximum and minimum diameters of isotopic particles, respectively. PSD of anisotropic particles can be connected to that of isotopic particles by substituting D_{eq} into D, as mentioned in Eq. (6.39). The optimal Fuller and EVF gradations for anisotropic particles are presented in the following equations

$$F_V\left(D_{eq}\right) = \frac{\sqrt{D_{eq}} - \sqrt{D_{\min eq}}}{\sqrt{D_{\max eq}} - \sqrt{D_{\min eq}}} \tag{6.40}$$

$$F_V\left(D_{\text{eq}}\right) = \frac{\text{In}\,D_{\text{eq}} - \text{In}\,D_{\min \text{eq}}}{\text{In}\,D_{\max \text{eq}} - \text{In}\,D_{\min \text{eq}}} \tag{6.41}$$

where $F_V\left(D_{\text{eq}}\right)$ should be derived to generate the number of polydispersed particles. The volume-based probability density function $f_V\left(D_{\text{eq}}\right)$ is determined in terms of D_{eq} according to the first-order derivative of $F_V\left(D_{\text{eq}}\right)$.

$$f_V\left(D_{eq}\right) = \frac{1}{\left(\sqrt{D_{\max eq}} - \sqrt{D_{\min eq}}\right)\sqrt{D_{eq}}} \tag{6.42}$$

$$f_V(D_{eq}) = \frac{1}{(\text{In} D_{\text{max eq}} - \text{In} D_{\text{min eq}}) D_{eq}} \quad (6.43)$$

The number dN of particles with various sizes in a range of $[D_{eq}, D_{eq} + dD_{eq}]$ is then calculated as:

$$dN = \frac{f_V(D_{eq}) dD_{eq}}{V(D_{eq})} \quad (6.44)$$

where $V(D_{eq})$ is the volume of anisotropic particles. Consequently, the number-based probability density function $f_N(D_{eq})$ for anisotropic particles is expressed as:

$$f_N(D_{eq}) = \frac{f_V(D_{eq})}{V(D_{eq})N} \quad (6.45)$$

and N is define by

$$N = \int_{D_{mineq}}^{D_{maxeq}} \frac{f_V(D_{eq})}{V(D_{eq})} dD_{eq} \quad (6.46)$$

By substituting Eqs. (6.40) and (6.41) into Eq. (6.46), $f_N(D_{eq})$ is given by:

$$f_N(D_{eq}) = \frac{-q}{\left(D_{\text{max } eq}^{-q} - D_{\text{min } eq}^{-q}\right) D_{eq}^{q+1}} \quad \begin{cases} q = 2.5 \rightarrow \text{Fuller gradation} \\ q = 3.0 \rightarrow \text{EVF gradation} \end{cases} \quad (6.47)$$

Moreover, the kth moment $\langle D_{eq}^k \rangle$ of area of $f_N(D_{eq})$ is given by

$$\langle D_{eq}^k \rangle = \int_{D_{\text{min } eq}}^{D_{\text{max } eq}} D_{eq}^k f_N(D_{eq}) dD_{eq} \quad (6.48)$$

By substituting Eq. (6.47) into Eq. (6.48), $\langle D_{eq} \rangle$, $\langle D_{eq}^2 \rangle$ and $\langle D_{eq}^3 \rangle$ are obtained to determine the volume fraction of interphase for an anisotropic polydispersed particle system. In addition, the parameters required to calculate $\text{Ø}_{\text{Interphase}}$ in term of $D_{\text{eq-min}}$, $D_{\text{eq-max}}$, $D_{\text{eq-mean}}$, H and $t_{\text{interphase}}$ can be then measured for each nanofiller.

6.3 Prediction of Elastic Moduli of PVA-Based Bionanocomposites

Elastic moduli of PVA/NBC bionanocomposites using Halpin–Tsai model based on two different orientation states were compared with experimental data, as depicted in Fig. 6.1a. In between, modelling parameters in Table 6.1 were utilised to substitute into Eqs. (6.1)–(6.6). Clearly, experimental data for bionanocomposite systems are in better agreement with Halpin–Tsai model (parallel) at low NBC nominal volume fractions ranging from 0.085 to 0.137 vol%. However, such data coincide more closely with Halpin–Tsai model (random) at the high NBC norminal volume fraction of 0.251 vol%. This phenomenon suggests that well-aligned NBC uniform dispersion may be prevalent at the low NBC contents, while heterogenous and randomly oriented NBCs become quite manifested due to large NBC agglomeration at the high NBC content levels. The general trend of experimental data in this study falls within the theoretical curves of Halpin–Tsai model (parallel) and Halpin–Tsai model (random) as upper and lower bounds, respectively, in good accordance with the previous work on PVA/PVA-g-GO nanocomposites [27].

Moreover, Mori–Tanaka model and the combination of laminate theory [12] and Mori–Tanaka model are also used to predict elastic moduli of PVA/NBC bionanocomposite by substituting the parameters in Table 6.1 into Eqs. (6.7)–(6.9), with associated results being presented in Fig. 6.1b. It is clearly seen that a similar trend can be identified when compared with predicted results based on Halpin–Tsai model. Apparently, experimental data of PVA/NBC bionanocomposites appear to be close to the estimated results by using Mori–Tanaka model (parallel) at the low NBC

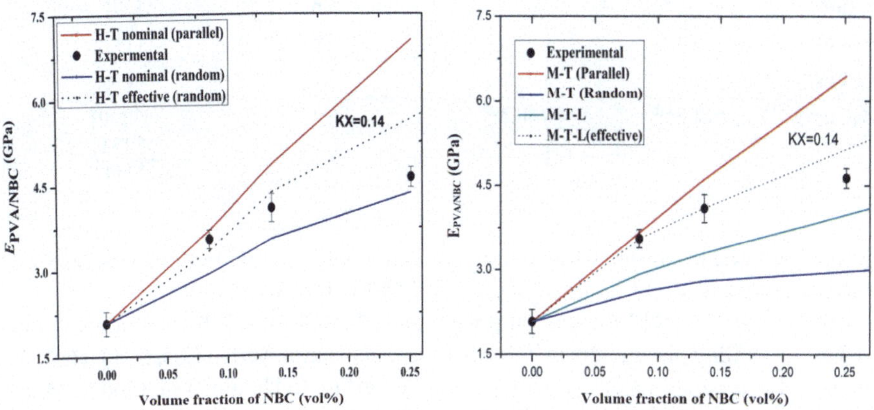

(a) (b)

Fig. 6.1 Prediction of Young's moduli of PVA/NBC bionanocomposites reinforced with well-aligned and randomly oriented NBCs represented by **a** Halpin–Tsai model (H-T), as well as **b** Mori–Tanaka model (M-T) and the combination of Mori–Tanaka model and laminate theory (M-T-L) using nominal and effective volume fractions of NBCs, respectively [25]

Table 6.1 Modelling parameters used for PVA bionanocomposite films [26]

Modelling parameter	Reference values			References
	PVA/NBC	PVA/HNT	PVA/Cloisite 30B	
E_m (GPa)	2.08	2.08	2.08	[29]
E_p (GPa)	84.5[a]	140[b]	178[c]	[a] [29] [b] [30] [c] [11]
l_p (nm)	–	208 (at 3 wt%) 410 (at 5 wt%) 900.6 (at 10 wt%)	109 (at 3 wt%) 206 (at 5 wt%) 560 (at 10 wt%	[29]
t_p (nm)	4.8 (at 3 wt%) 11.2 (at 5 wt%) 28.4 (at 10 wt%)	–	4.6 (at 3 wt%) 15.8 (at 5 wt%) 41.6 (at 10 wt%)	[29]
d (nm)	27.6 (at 3 wt%) 78.4 (at 5 wt%) 232 (at 10 wt%)	35.2 (at 3 wt%) 66.5 (at 5 wt%) 85 (at 10 wt%)	–	[29]
ρ_m (g cm^{-3})	0.39	0.39	0.39	[29]
ρ_p (g cm^{-3})	1.3	1.8	1.98	Material data sheet
t (nm)	~58	~7.3	~10.4	[29]
R_g (nm)	10	10	10	[11]
ν_m	0.4	0.4	0.4	[13]
W''_p	–	–	0.66	[11]
d'_1 (nm)	–	–	1.87	[29]
d_2 (nm)	–	–	3.6	[29]
N (nm)	–	–	5.6	[29]
h (nm)			0.98	[11]
ρ''_p (g cm^{-3})			3.1	[11]
A_T (m^2 g^{-1})	624.81[d]	40.3[e]	658[f]	[d] [29] [e] [31] [f] [11]

nominal volume fractions, while better agreement with the predictions was obtained with the combination of laminate theory and Mori–Tanaka model.

In case of PVA/HNT bionanocomposites and PVA/Cloisite 30B clay bionanocomposites, modelling parameters in Table 6.1 were also employed in Eqs. (6.1)–(6.9) for the modulus predictions of nanocomposites according to Halpin–Tsai model, Mori–Tanaka model, as well as the combination of laminate theory [12] and Mori–Tanaka model, as illustrated in Figs. 6.2 and 6.3. In both nanocomposite systems, experimental data did not have good accordance with predicted results at low nominal volume fractions of HNTs and Cloisite 30B clays. The deviation between experimental data and predicted values is related to the concept that Halpin–Tsai model and Mori–Tanaka model are categorised as micromechanical models that are more

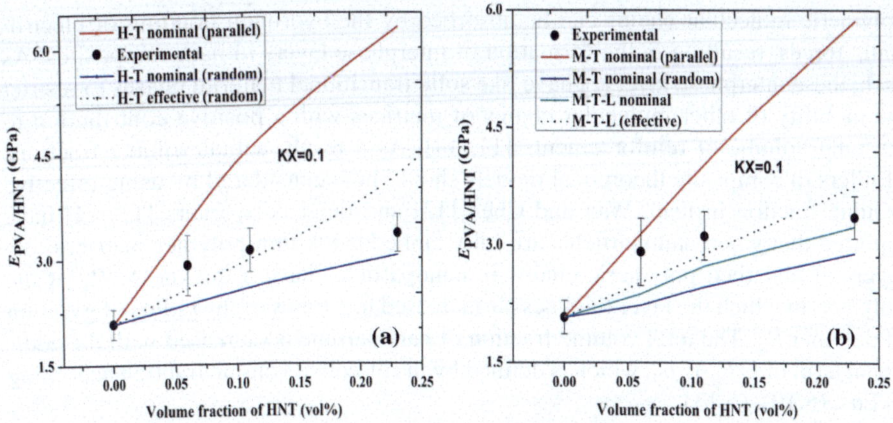

Fig. 6.2 Prediction of Young's moduli of PVA/HNT bionanocomposites reinforced with well-aligned and randomly oriented HNTs represented by **a** Halpin–Tsai model (H-T), as well as **b** Mori–Tanaka model (M-T) and the combination of Mori–Tanaka model and laminate theory (M-T-L) using nominal and effective volume fractions of HNTs, respectively [26]

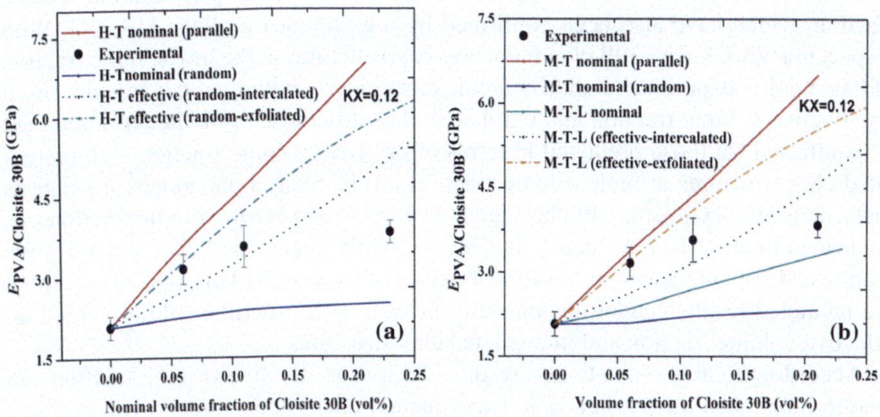

Fig. 6.3 Prediction of Young's moduli of PVA/Cloisite 30B clay bionanocomposites reinforced with well-aligned and randomly oriented clays represented by **a** Halpin–Tsai model (H-T), as well as **b** Mori–Tanaka model (M-T) and the combination of Mori–Tanaka model and laminate theory (M-T-L) using nominal and effective volume fractions of clays, respectively [26]

feasible to conventional composites rather than nanocomposites. Furthermore, the major drawback in using Halpin–Tsai model and Mori–Tanaka model lies in their common assumption that both fillers and polymer matrices in nanocomposites are linearly elastic and isotropic materials with uniform filler distribution, as well as the neglect of particle–particle interaction [28]. In a nanocomposite system, isotropic nanoparticles generally have large surface areas, on which substantial amounts of polymeric molecular chains can be absorbed by the hydrogen bonding or electrostatic forces, resulting in the formation of interphase layers in nanocomposites. As such, these interphase layers behave like solid transitional material phases to restrict the mobility of other molecular chains of matrices with a positive contribution to the total volume of reinforcements [11, 14]. As a result, actual volume fractions of fillers in composite theoretical models should be reconsidered by using effective volume fraction instead. Wan and Chen [13] and other co-workers [11, 14] have reported that when nanoparticles are fully embedded within polymer matrices, the layers of absorbed polymeric chains in nanoparticles have a thickness xR_g of the polymer, in which the layer thickness is increased to x times of the radius of gyration of polymer R_g. The total volume fraction of nanoparticle is increased with the additional term of $kxR_gA_T\rho_p$, which is defined by the effective volume fraction according to Eqs. (6.10)–(6.13).

The parameters of PVA-based bionanocomposite films in Table 6.1 were used to estimate effective volume fraction, and subsequently predicted results obtained from Halpin–Tsai model and Mori–Tanaka model were determined in terms of effective volume fraction of nanofillers, as illustrated in Figs. 6.1, 6.2 and 6.3. The predictions on the basis of effective volume fraction of nanofillers appear to be in better agreement with experimental data, as opposed to those corresponding to nominal volume fraction, which have also been confirmed by the previous findings [13, 14]. With respect to PVA/Cloisite 30B clay bionanocomposite films at the low volume fraction of 0.085 vol%, experimental data coincide more closely with predicted results based on effective volume fraction and exfoliated clay structures, while better agreement is manifested for those predicted in terms of effective volume fraction and intercalated clay structures at high volume fraction levels. Such a phenomenon suggests that a majority of Cloisite 30B clays tend to be exfoliated at low volume fractions, as confirmed by the AFM evaluation in Chap. 4, while intercalated clay structures are manifested with increasing the volume fraction of Cloisite 30B clays. Consequently, the predicted results cannot unanimously fit well with experimental data based on effective volume fraction and intercalated clay structures.

According to above-mentioned results, the use of effective volume fraction can lead to more accurate prediction to experimental data in comparison with nominal volume fraction in PVA-based bionanocomposites. However, the layer thickness of polymeric molecules absorbed on nanofiller surfaces (kx) is even in the same grades of polymers with different molecular structures and polarities. As noted earlier in Chap. 5, interphase features do not possess uniform layer thickness, and their dimensions and properties vary greatly when embedded with heterogeneous nanoparticles by altering particle sizes/dimensions, as well as nanomechanical properties associated with particle/filler interactions. As a result, it is suggested that a new and simple formula in Eq. (6.14) should be utilised to calculate nanofiller volume fraction from both interphase volume and volume fraction, in which interphase volume fraction $\emptyset_{\text{Interphase}}$ is estimated in two different systems, namely uniform particle monodispersion and particle polydispersion. Modelling parameters in Table 6.2 are substituted into Eqs. (6.37) and (6.29), respectively, to estimate associated $\emptyset_{\text{Interphase}}$ values in

Table 6.2 Modelling parameters measured for calculating $Ø_{\text{Interphase}}$ in PVA-based bionanocomposites [26]

Modelling parameter	Nanofiller content (wt%)	PVA/NBC	PVA/HNT[b]	PVA/Cloisite 30B clay
$D_{\text{eq-min}}$ (nm)	3	14	64.5	34.1
	5	32.5	168.5	54.3
	10	66	245.7	141.6
$D_{\text{eq-max}}$ (nm)	3	74.3	836	436
	5	1017	2460	1836.2
	10	2013	4645	3141
$D_{\text{eq-mean}}$ (nm)	3	27	208	109
	5	75.5	410	206
	10	232	900.6	560
$t^{a}_{\text{interphase}}$ (nm)	5	15.3	8.6	10.7

[a] $t_{\text{interphase}}$ is the average interphase thickness for the model development based on the assumption that the interphase generated between nanoparticles and polymer matrices was typically considered at the filler content of 5 wt% in PVA bionanocomposites

[b] $D_{\text{eq-min}}$, $D_{\text{eq-max}}$ and $D_{\text{eq-mean}}$ are the maximum, minimum and mean equivalent diameters of anisotropic particles, which represented the maximum, minimum and mean equivalent lengths of HNTs in this case, respectively, in PVA/HNT bionanocomposites according to Eqs. (6.38)–(6.48)

PVA bionanocomposite systems in this study. Besides, interphase volume can be calculated according to Eqs. (5.9) and (5.10) in Chap. 5 with the relevant results being used for calculating volume fractions of nanofillers. Such recalculated volume fractions of nanoparticles in relation to interphase effect were incorporated into Halpin–Tsai model and Mori–Tanaka model accordingly along with the predicted results presented for elastic moduli of PVA-based bionanocomposites in Figs. 6.4, 6.5 and 6.6. In a consistent manner, all E_c values in our modelling work by considering interphase effect demonstrate a much closer relationship with experimental data, as opposed to corresponding theoretical models using both nominal and effective volume fractions without interphase. More evidently, our proposed models in a monodispersed particle system appear to maintain relatively good agreement with experimental data at a low volume fraction up to 0.0604 and 0.061 vol% for HNTs and Cloisite 30B clays in contrast with those in a polydispersed particle system with both Fuller and EVF particulate gradations, but vice versa beyond 0.1109 and 0.1105 vol%, respectively. This trend is easily understood since NBCs, HNTs and Cloisite 30B clays are expected to be more uniformly dispersed in PVA bionanocomposites in relatively similar sizes, as validated by our morphological structure results, while typical nanoparticle agglomeration is evidently observed to have a better correlation with a polydispersed particle system. In terms of particulate gradation effect, those predictions based on the Fuller gradation appear to be slightly in better agreement with experimental data, possibly resulting from the theoretical assumption of finer particle dispersion taking place in the EVF counterpart at the same nominal volume

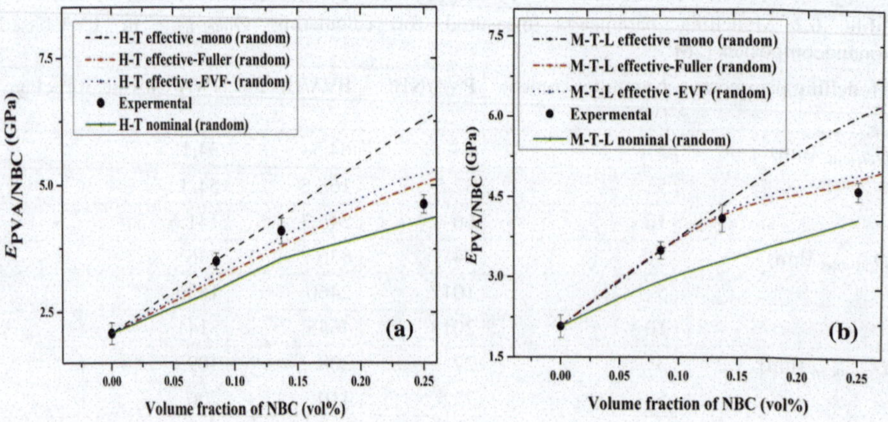

Fig. 6.4 Elastic moduli of PVA/NBC bionanocomposite films predicted by **a** Halpin–Tsai model (H-T) and **b** the combination of Mori–Tanka model and laminate theory (M-T-L). Four categories are considered based on nominal volume fraction of NBCs, effective volume fraction of NBCs in a monodispersed particle system, effective volume fraction of NBCs in a polydispersed particle system with Fuller particular gradation, as well as effective volume fraction of NBCs in a polydispersed particle system with EVF particulate gradation

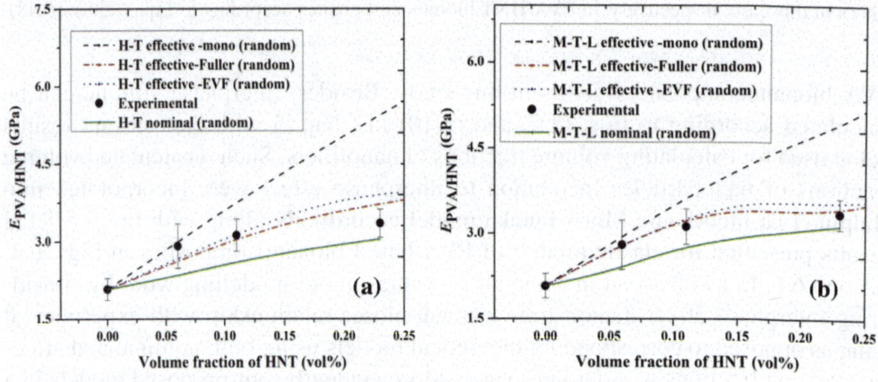

Fig. 6.5 Elastic moduli of PVA/HNT bionanocomposites predicted by **a** Halpin–Tsai model (H-T) and **b** the combination of Mori–Tanka model and laminate theory (M-T-L). Four categories are considered based on nominal volume fraction of HNTs, effective volume fraction of HNTs in a monodispersed particle system, effective volume fraction of HNTs in a polydispersed particle system with Fuller particular gradation, as well as effective volume fraction of HNTs in a polydispersed particle system with EVF particulate gradation [26]

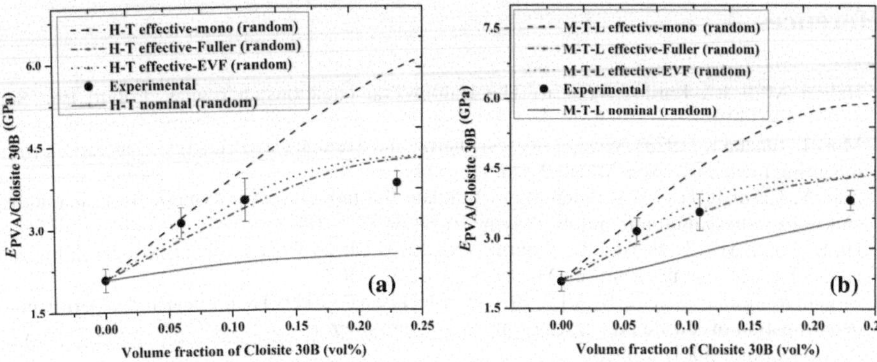

Fig. 6.6 Elastic moduli of PVA/Cloisite 30B clay bionanocomposites predicted by **a** Halpin–Tsai model (H-T) and **b** the combination of Mori–Tanka model and laminate theory (M-T-L). Four categories are considered based on nominal volume fraction of clays, effective volume fraction of clays in a monodispersed particle system, effective volume fraction of clays in a polydispersed particle system with Fuller particular gradation, as well as effective volume fraction of clays in a polydispersed particle system with EVF particulate gradation [26]

fraction. As a result, larger specific surface areas of particles take place in case of EVF gradation, further leading to the increases in $\emptyset_{\mathrm{Interphase}}$ and volume fraction of nanoparticles.

6.4 Summary

In this chapter, the following key points are summarised:

- Halpin–Tsai model and the combination of Mori–Tanaka model and laminate theory were successfully employed to predict elastic moduli of PVA-based bionanocomposites.
- Effective volume fraction of reinforcing phases was introduced in conventional composite theory with the consideration of interphase for the reinforcement effect, resulting in better agreement with experimental data, as opposed to those based on nominal volume fraction.
- For the first time, we have also developed a theoretical approach to calculate effective volume fraction of nanoparticles based on 3D interphase dimensions and volume fraction, which is allowed to be used for studying the effect of experimentally measured interphase on the elastic moduli of PVA bionanocomposite films. Besides, the application of effective volume fraction of reinforcing phases in Halpin–Tsai model and the combination of Mori–Tanaka model and laminate theory in a polydispersed particle system with Fuller and EVF particulate gradations demonstrates much better agreement with experimental data, as compared with those using nominal volume fraction of nanoparticles.

References

1. Halpin Affdl JC, Kardos J (1976) The Halpin-Tsai equations: a review. Polym Eng Sci 16(5):344–352
2. Mori T, Tanaka K (1973) Average stress in matrix and average elastic energy of materials with misfitting inclusions. Acta Metall 21(5):571–574
3. Zare Y, Garmabi H (2014) Attempts to simulate the modulus of polymer/carbon nanotube nanocomposites and future trends. Polym Rev 54(3):377–400
4. Hu K, Gupta MK, Kulkarni DD, Tsukruk VV (2013) Ultra-robust graphene oxide-silk fibroin nanocomposite membranes. Adv Mater 25(16):2301–2307
5. Sisakht Mohsen R, Saied NK, Ali Z, Hosein EM, Hasan P (2009) Theoretical and experimental determination of tensile properties of nanosized and micron-sized $CaCO_3$/PA66 composites. Polym Compos 30(3):274–280
6. Zare Y (2016) Development of simplified Tandon-Weng solutions of Mori-Tanaka theory for Young's modulus of polymer nanocomposites considering the interphase. J Appl Polym Sci 133(33):43816
7. Zare Y (2016) The roles of nanoparticles accumulation and interphase properties in properties of polymer particulate nanocomposites by a multi-step methodology. Compos Part A: Appl Sci Manuf 91:127–132
8. Dwivedi H, Mathur RB, Dhami TL, Bahl OP, Monthioux M, Sharma SP (2006) Evidence for the benefit of adding a carbon interphase in an all-carbon composite. Carbon 44(4):699–709
9. Mousa M, Dong Y (2018) Novel three-dimensional interphase characterisation of polymer nanocomposites using nanoscaled topography. Nanotechnology 29(38):385701
10. Tandon GP, Weng GJ (1984) The effect of aspect ratio of inclusions on the elastic properties of unidirectionally aligned composites. Polym Compos 5(4):327–333
11. Chen B, Evans JR (2006) Nominal and effective volume fractions in polymer-clay nanocomposites. Macromolecules 39(5):1790–1796
12. Van Es MA (2001). Polymer-clay nanocomposites: the importance of particle dimensions. Ph.D. thesis, Delft University of Technology, Netherlands
13. Wan C, Chen B (2012) Reinforcement and interphase of polymer/graphene oxide nanocomposites. J Mater Chem 22(8):3637–3646
14. Wan C, Frydrych M, Chen B (2011) Strong and bioactive gelatin–graphene oxide nanocomposites. Sof Matter 7(13):6159–6166
15. Buenviaje C, Ge S, Rafailovich M, Sokolov J, Drake JM, Overney RM (1999) Confined flow in polymer films at interfaces. Langmuir 15(19):6446–6450
16. Lu B, Torquato S (1992) Nearest-surface distribution functions for polydispersed particle systems. Phys Rev A 45(8):5530
17. Torquato S (2002) Random heterogeneous materials: microstructure and macroscopic properties. Springer, New York
18. Mansoori G, Carnahan NF, Starling KE, Leland TW Jr (1971) Equilibrium thermodynamic properties of the mixture of hard spheres. J Chem Phys 54(4):1523–1525
19. Underwood EE (1969) Stereology, or the quantitative evaluation of microstructures. J Microsc 89(2):161–180
20. Xu WX, Chen HS (2013) Analytical and modeling investigations of volume fraction of interfacial layers around ellipsoidal aggregate particles in multiphase materials. Model Simul Mat Sci Eng 21(1):015005
21. Xu W, Chen W, Chen H (2014) Modeling of soft interfacial volume fraction in composite materials with complex convex particles. J Chem Phys 140(3):034704
22. Xu WX, Chen HS (2013) Numerical investigation of effect of particle shape and particle size distribution on fresh cement paste microstructure via random sequential packing of dodecahedral cement particles. Comput Struct 114–115:35–45
23. Xu W, Duan Q, Ma H, Chen W, Chen H (2015) Interfacial effect on physical properties of composite media: Interfacial volume fraction with non-spherical hard-core-soft-shell-structured particles. Sci Rep 5:16003

24. Gao Y, Schutter GD, Ye G (2013) Micro-and meso-scale pore structure in mortar in relation to aggregate content. Cement Concrete Res 52:149–160
25. Mousa M, Dong Y (2017) Strong poly (vinyl alcohol)(PVA)/bamboo charcoal (BC) nanocomposite films with particle size effect. ACS Sustain Chem Eng 6(1):467–479
26. Mousa M, Dong Y (2020) Towards sophisticated 3D interphase modelling of advanced bionanocomposites via atomic force microscopy. J Nanomater Article ID 4526108
27. Cheng HKF, Sahoo NG, Tan YP, Pan Y, Bao H, Li L, Chan SH, Zhao J (2012) Poly(vinyl alcohol) nanocomposites filled with poly(vinyl alcohol)-grafted graphene oxide. ACS Appl Mater Interfaces 4:2387–2394
28. Fornes TD, Paul DR (2003) Modeling properties of nylon 6/clay nanocomposites using composite theories. Polymer 44(17):4993–5013
29. Mousa M (2018) Experimental characterisation and modelling of sustainable multiscaled bionanocomposites. Ph.D. thesis, Curtin University, Perth, Australia
30. Lecouvet B, Horion J, D'haese C, Bailly C, Nysten B (2013) Elastic modulus of halloysite nanotubes. Nanotechnology 24(10):105704
31. Yuan P, Southon PD, Liu Z, Green ME, Hook JM, Antill SJ, Kepert CJ (2008) Functionalization of halloysite clay nanotubes by grafting with γ-aminopropyltriethoxysilane. J Phys Chem C 112(40):15742–15751

Chapter 7
Applications of Bionanocomposite Materials

Abstract The development of bionanocomposites reinforced with different nanofillers such as organically modified layered silicates (OMLs), bamboo charcoals (BCs) and halloysite nanotubes (HNTs) has drawn great attention by research and industrial sectors owing to their unique material merits like excellent biodegradability, natural abundance, cost-effectiveness and ecofriendliness. This chapter provides a brief review on the potential applications of typical bionanocomposites in the fields of electronics, medical devices and material packaging, etc.

Keywords Bionanocomposites · Nanofillers · Electronics · Medical devices · Material packaging

7.1 Electronic and Sensor Applications

When biopolymers are reinforced with nanofillers, their resulting bionanocomposites not only enhance mechanical, thermal and barrier properties, but also possess electromagnetic shielding, good electrical, optical and magnetic properties [1]. Such multi-functional properties of bionanocomposites benefit from an increasingly high environmental concern of replacing conventional non-degradable composites with bionanocomposites, particularly in manufacturing electric products with good recyclability and renewability. Most current polarising films in LCD screens are made of iodine-doped PVA. Novel electroactive shape-memory materials based on PVA/multi-walled carbon nanotube (MWNT) nanocomposites possess higher thermal and electrical properties [2]. On the other hand, transparent and flexible bionanocomposites with the incorporation of nanocellulose are very useful in various applications ranging from displays, organic light-emitting diode, solar cells and roll-to-roll technology. Electronic deposition devices depend on the continuous deposition of various functional components to be used, among which metal wiring and active gas barrier films are good representatives for roll-to-roll technology [3]. Whereas, the major drawback lies in a higher coefficient of thermal expansion (CTE) of plastic materials [3], which can be overcome by reinforcing bacterial cellulose as nanofillers resulting in the reduction in CTE. When such bionanocomposites are produced, they can act as potential substrates used for the roll-to-roll fabrication [4].

M. Mousa and Y. Dong, *Multiscaled PVA Bionanocomposite Films*,
https://doi.org/10.1007/978-981-15-8771-9_7

7.2 Medical Applications

Natural abundance, adaptability and environmental friendliness make bionanocomposites a suitable material candidate for a wide range of medical applications such as the regeneration of natural tissue structures. In this area, bionanocomposites can be fabricated to have the same morphology as that of extracellular matrix (ECM) including protein, polysaccharides and inorganic matters [5]. For bone regeneration of clinical orthopaedics, the selection of bionanocomposites such as polymer scaffolds and hydroxyapatites (HAPs) with biodegradable and bioresorbable structures is critical to stimulate the bone formation at the implantation sites [6] More importantly, sufficient porosity is required for the structures of such bionanocomposites so that new tissue growth can be facilitated like porous scaffolds based on HAP/collagen nanocomposites in order to repair bony defects in animal models [7]. It is worth mentioning that PVA/HNT bionanocomposite films have also been utilised in tissue engineering with good compatibility to osteoblast and fibroblast cells [8].

In a similar manner, chitosan/HNT nanocomposite scaffolds exhibit highly porous structures with high compressive strength and compressive modulus for the development of mouse fibroblasts [9]. No adverse effect is evidently shown for electrospun poly(lactic-co-glycolic acid) (PLGA)/HNT bionanocomposites with high biocompatibility and cell proliferation when incorporated with HNTs [10]. The potential of using bionanocomposites in drug delivery has been widely reported in the previous studies [11, 12]. Nanocomposites reinforced with HNTs become a promising drug carrier to be loaded in different forms of powders, suspensions and fibrous scaffolds [13]. A slow release rate occurs when the lumen and outer surfaces of HNTs are loaded with diltiazem hydrochloride, 5-aminosalicylic acid and propanol hydrochloride [13]. However, HNT loading capacity is limited because the lumen volume fraction of HNTs cannot exceed 10 vol% [14]. The increase in loading capacity of HNTs can be achieved by etching alumina from HNT inner surfaces. Additionally, the inclusion of coated HNTs into biopolymers such as PVA [15], poly (ethyleneimine) (PEI) [16] and chitosan [17] has also been shown to significantly retard a drug release rate, as opposed to those uncoated counterparts.

7.3 Packaging Applications

In general, biopolymers have major drawbacks such as narrow processing window, poor gas/water barrier properties, as well as weak mechanical properties when compared with those of conventional synthetic polymers, which can be overcome by the inclusion of nanofillers into neat biopolymers to produce bionanocomposite materials. The characteristic permeability is one of the most important factors in food packaging industries when suitable materials are selected [18].

The incorporation of nanofillers like nanoclays into biopolymers yields a remarkable improvement in barrier properties of bionanocomposites [1]. As a fundamental principle, the transfer of molecules in polymer/clay nanocomposites requires a much longer path rather than directly diffuse into neat polymers. As such, more tortuous diffusive paths can be generated in a nanocomposite system resulting in a delay in the transfer of molecules [18]. Because of good resistance to most organic compounds and solvents, PVA and PVA bionanocomposites have been widely used in material packaging industries so that packed products can be protected from the secondary contamination by printing links [19]. Furthermore, the permeability is prone to the reduction with increasing the nanofiller content. Aloui et al. [20] reported that the addition of 3 wt% HNTs and 5 wt% cellulose nanocrystals (CNCs) within PVA matrices in PVA bionanocomposites decreased both water and oxygen permeabilities. Similarly, Strawhecker and Manias [21] found that with the inclusion of 4–6 wt% nanofillers, water vapour permeability of PVA/Na^+ montmorillonite (MMT) nanocomposite films was decreased by 40% when compared with that of neat PVA. This result arises from a mix of exfoliated/intercalated clay structures, which generates a tortuous diffusion path through nanocomposite films. On the other hand, the addition of graphene oxide (GO) particles within PVA matrices drastically reduced the oxygen permeability owing to the exfoliation of GOs with a resulting tortuous path again for oxygen molecules [22]. Kim et al. [23] revealed that the permeability of PVA/GO nanocomposite films was found to be decreased by 20% with the inclusion of only 0.3 wt% GOs relative to that of pure PVA films. Polyhydroxyalkanoate (PHA) has impressive properties such as hydrophobic nature and water vapour barrier properties, which are similar to those for polyethylene (PE) coating applications. The drawback of PHA lies in its poor gas barrier properties and narrow processing window, thus limiting its applications in food packaging. The addition of nanoclays in PHA was found to result in much lower oxygen and water permeabilities [24]. On the other hand, starch as one of the most popular biopolymers has high hydrophilicity, and thus limits its usage in material packaging applications. The inclusion of nanofillers into thermoplastic starch (TPS) is regarded as a useful method to overcome such a drawback with success. Park et al. [25] presented that the water vapour transition rate of TPS/MMT nanocomposites was reduced by a half, as opposed to that of pure TPS, owing to the tortuous pathway generated by embedded MMTs. On the contrary, Change et al. [26] evaluated the effect of chitosan nanoparticles (CNs) on the reduction in barrier properties of TPS because CN shape could not induce a sufficient tortuous path with a certain limitation in contrast with platelet-like MMTs. Overall, the applications of bionanocomposites may yield enormous applications in material packaging with sophisticated development, important innovation and pronounced cost-effectiveness.

7.4 Summary

Increasing environmental concerns due to the use of non-degradable polymers have motivated the researchers and material developers to find alternative polymeric materials. Biodegradable polymers are considered as a suitable material candidate for conventionally synthetic polymers in order to reduce energy consumption and gas emission. Key biopolymers applications with their properties in relation to electronics and material packaging have been particularly reviewed in this chapter. Moreover, the material performance of bionanocomposite films developed in our study can be improved with the following recommendations:

- Our results indicate that BCs are effective nanofillers for the property enhancement of PVA nanocomposites due to their existing mechanical bonding with PVA matrices based on typical highly porous BC structures. The reinforcement efficiency of BCs could be further enhanced by their surface modification to use with different hydrophilic and hydrophobic polymers in the development of advanced nanocomposites, which can directly benefit from enlarging pore diameters and the supply of more surface functional groups. As such, their interaction and miscibility can be promoted with different types of polymers as matrices in nanocomposite systems.
- BCs in possession of excellent absorption capacity could be employed for retaining food and inhibiting the growth of bacteria, particularly for the fabrication of PVA bionanocomposite films with their potential applications in packaging applications.
- On the other hand, BCs have large surface area, absorbance efficiency and biocompatibility, and can be considered as a suitable carrier candidate for drug delivery with their ability to improve the human microcirculation system. The similar case can also apply to PVA/HNT nanocomposites as HNTs are well known to be effective carriers for drug delivery.
- Although micromechanical models used in this study are developed to evaluate the impact of 3D interphase dimension, interphase volume, aspect ratio and content of nanofillers, as well as nanofiller dispersion states with much better predictions for tensile moduli of PVA bionanocomposites, interphase modulus should also be considered for better understanding the overall macroscopic material behaviour of bionanocomposites.
- PVA bionanocomposites were successfully manufactured by a solution casting method leading to strong PVA/NBC bionanocomposites, and good PVA/HNT bionanocomposites and PVA/Cloisite 30B clay bionanocomposites. A layer-by-layer method is also suggested in order to achieve more desirable properties of bionanocomposite thin films. In such a process, deposition layers of polymer and nanofiller suspension can be placed on the substrates based on other different technologies such as dipping, dewetting, roll-to-roll process and centrifugation for the assemblies to generate more desired films on the substrate.

References

1. Mousa MH, Dong Y, Davies IJ (2016) Recent advances in bionanocomposites: preparation, properties, and applications. Int J Polym Mater Polym Biomater 65(5):225–254
2. Du F-P, Ye EZ, Yang W, Shen TH, Tang CY, Xie XL, Zhou XP, Law WC (2015) Electroactive shape memory polymer based on optimized multi-walled carbon nanotubes/polyvinyl alcohol nanocomposites. Compos Part B: Eng 68:170–175
3. Forrest SR (2004) The path to ubiquitous and low-cost organic electronic appliances on plastic. Nature 428(6986):911–918
4. Nogi M, Yano H (2008) Transparent nanocomposites based on cellulose produced by bacteria offer potential innovation in the electronics device industry. Adv Mater 20(10):1849–1852
5. McCullen SD, Ramaswamy S, Clarke LI, Gorga RE (2009) Nanofibrous composites for tissue engineering applications. Wiley Interdiscip Rev Nanomed Nanobiotechnol 1(4):369–390
6. Hule RA, Pochan DJ (2007) Polymer nanocomposites for biomedical applications. MRS Bull 32(4):354–358
7. Yunoki S, Ikoma T, Monkawa A, Ohta K, Kikuchi M, Sotome S, Shinomiya K, Tanaka J (2006) Control of pore structure and mechanical property in hydroxyapatite/collagen composite using unidirectional ice growth. Mater Lett 60(8):999–1002
8. Zhou WY, Guo B, Liu M, Liao R, Rabie AB, Jia D (2010) Poly (vinyl alcohol)/halloysite nanotubes bionanocomposite films: Properties and in vitro osteoblasts and fibroblasts response. J Biomed Mater Res A 93(4):1574–1587
9. Liu M, Wu C, Jiao Y, Xiong S, Zhou C (2013) Chitosan–halloysite nanotubes nanocomposite scaffolds for tissue engineering. J Mater Chem B 1(15):2078–2089
10. Qi R, Guo R, Shen M, Cao X, Zhang L, Xu J, Yu J, Shi X (2010) Electrospun poly (lactic-co-glycolic acid)/halloysite nanotube composite nanofibers for drug encapsulation and sustained release. J Mater Chem 20(47):10622–10629
11. Levis S, Deasy P (2002) Characterisation of halloysite for use as a microtubular drug delivery system. Int J Pharm 243(1–2):125–134
12. Aguzzi C, Viseras C, Cerezo P, Salcedo I, Sánchez-Espejo R, Valenzuela C (2013) Release kinetics of 5-aminosalicylic acid from halloysite. Colloid Surf B: Biointerf 105:75–80
13. Liu M, Jia Z, Jia D, Zhou C (2014) Recent advance in research on halloysite nanotubes-polymer nanocomposite. Progr Polym Sci 39(8):1498–1525
14. Hasook A, Muramatsu H, Tanoue S, Iemoto Y, Unryu T (2008) Preparation of nanocomposites by melt compounding polylactic acid/polyamide 12/organoclay at different screw rotating speeds using a twin screw extruder. Polym Compos 29(1):1–8
15. Ghebaur A, Garea SA, Lovu H (2012) New polymer-halloysite hybrid materials-potential controlled drug release system. Int J Pharm 436(1–2):568–573
16. Levis SR, Deasy PB (2003) Use of coated microtubular halloysite for the sustained release of diltiazem hydrochloride on propranolol hydrochloride. Int J Pharm 253(1–2):145–157
17. Kelly HM, Deasy PB, Ziaka E, Claffey N (2004) Formulation and preliminary in vivo dog studies of a novel drug delivery system for the treatment of periodontitis. Int J Pharm 274(1–2):167–183
18. Ray SS, Okamoto M (2003) Polymer/layered silicate nanocomposites: a review from preparation to processing. Progr Polym Sci 28(11):1539–1641
19. Tang X, Alavi S (2011) Recent advances in starch, polyvinyl alcohol based polymer blends, nanocomposites and their biodegradability. Carbohydr Polym 85(1):7–16
20. Aloui H, Khwaldia K, Hamdi M, Fortunati E, Kenny JM, Buonocore GG, Lavorgna M (2016) Synergistic effect of halloysite and cellulose nanocrystals on the functional properties of PVA based nanocomposites. ACS Sustain Chem Eng 4(3):794–800
21. Strawhecker K, Manias E (2000) Structure and properties of poly (vinyl alcohol)/Na^+ montmorillonite nanocomposites. Chem Mater 12(10):2943–2949
22. Loryuenyong V, Saewong C, Aranchaiya C, Buasri A (2015) The improvement in mechanical and barrier properties of poly (vinyl alcohol)/graphene oxide packaging films. Packag Technol Sci 28(11):939–947

23. Kim HM, Lee JK, Lee HS (2011) Transparent and high gas barrier films based on poly (vinyl alcohol)/graphene oxide composites. Thin Solid Films 519(22):7766–7771
24. Sanchez-Garcia M, Lagaron J (2010) Novel clay-based nanobiocomposites of biopolyesters with synergistic barrier to UV light, gas, and vapour. J Appl Polym Sci 118(1):188–199
25. Park H-M, Lee WK, Park CY, Cho WJ, Ha CS (2003) Environmentally friendly polymer hybrids Part I Mechanical, thermal, and barrier properties of thermoplastic starch/clay nanocomposites. J Mater Sci 38(5):909–915
26. Chang PR, Jian R, Yu J, Ma X (2010) Fabrication and characterisation of chitosan nanoparticles/plasticised-starch composites. Food Chem 120(3):736–740

Appendix
Interphase Dimension Measurement Via AFM

PVA/NBC Bionanocomposite Films

According to our previous work [1, 2], NBCs are regarded as anisotropic particles resulting in the interaction with polymer matrices to generate interphase. Interphase could also be anisotropic between PVA/NBC phases. The variation of interphase dimensions leads to different nanomechanical properties to influence overall material performance of bionanocomposites. A simple approach to measure 3D interphase dimensions to determine interphase surface area and interphase volume based on nanomechanical properties has been developed as follows:

- At the first step, interphase dimensions in term of interphase width $W_{\mathrm{Interphase}}$, interphase length $L_{\mathrm{Interphase}}$ and interphase height $H_{\mathrm{Interphase}}$ within PVA/NBC phases were detected. Such dimensions could be determined experimentally by using PFQNM with the typical features of distinguishing the interphase between nanoparticles and polymer matrices based on different nanomechanical properties. Therefore, PVA/NBC phases were first scanned transversely by PFQNM at a scanning interval of 10 nm, in which the interphase width $W_{i\mathrm{Interphase}}$ along the ith transverse plane ($i = 1, 2, 3, \ldots$) is equal to the summation of individual NBC width $W_{i\mathrm{NBC}}$ and its corresponding transverse interphase thicknesses t_{wi1} and t_{wi2}, as illustrated in Fig. A1, namely

$$W_{i\mathrm{Interphase}} = W_{i\mathrm{NBC}} + t_{\mathrm{Wi1}} + t_{\mathrm{Wi2}} \tag{A1}$$

The same applies to the further AFM measurements of interphase length ($L_{j\mathrm{Interphase}}$) and interphase height ($H_{k\mathrm{Interphase}}$) with similar equations being used as follows:

$$L_{j\mathrm{Interphase}} = L_{j\mathrm{NBC}} + t_{\mathrm{Lj1}} + t_{\mathrm{Lj2}} \tag{A2}$$

$$H_{k\mathrm{Interphase}} = H_{k\mathrm{NBC}} + t_{\mathrm{Hk1}} + t_{\mathrm{Hk2}} \tag{A3}$$

M. Mousa and Y. Dong, *Multiscaled PVA Bionanocomposite Films*,
https://doi.org/10.1007/978-981-15-8771-9

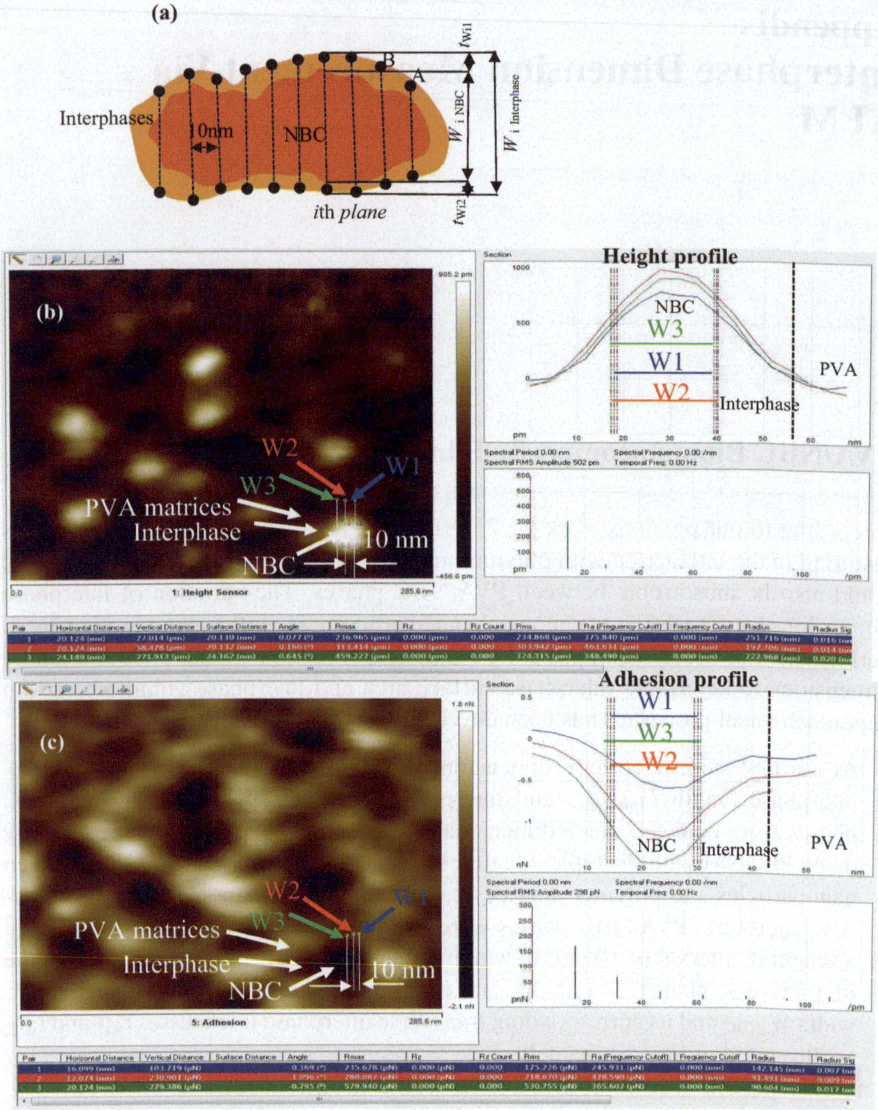

Fig. A1 **a** Schematic diagrams of cross-sectional scanning of PVA/NBC interphase in PVA/NBC bionanocomposites along transverse plane with **b** height profile and **c** adhesion profile [1]

where the interphase length $L_{j\text{Interphase}}$ along the jth longitudinal plane ($j = 1, 2, 3, \ldots$) is equal to the summation of individual NBC length $L_{j\text{NBC}}$ and its corresponding longitudinal interphase thicknesses $t_{\text{L}j1}$ and $t_{\text{L}j2}$, Fig. A2a. Furthermore, as shown in Fig. A2b, the interphase height $H_{k\text{Interphase}}$ along the kth height plane ($k = 1, 2,$

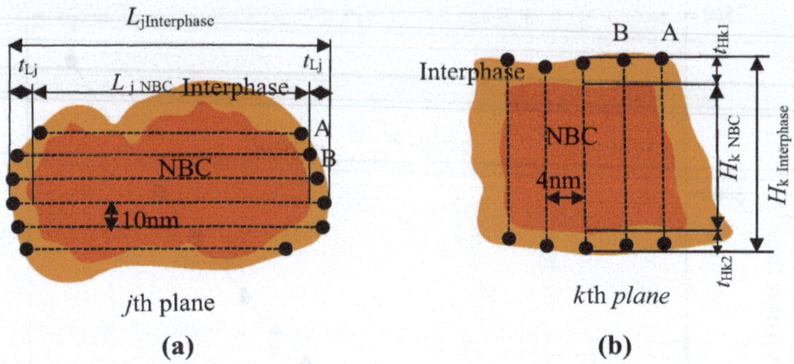

Fig. A2 Schematic diagrams of cross-sectional scanning of PVA/NBC interphases in PVA/NBC bionanocomposites along, **a** longitudinal plane in a top view and **b** height plane in a side view [1]

3, …) can be given by adding up individual NBC height $H_{k\text{NBC}}$, height interphase thicknesses t_{Hk1} and t_{Hk2}.

- The next step is to employ these dimensions in calculating interphase surface areas. In PVA/NBC bionanocomposites, surface area calculations of NBCs and interphases should take into consideration on the basis of the fundamental concept of particle shapes of anisotropic NBCs. According to Behmer and Hawkins [3], surface area (SA_p) of anisotropic particles could be calculated in the following equation:

$$\text{SA}_\text{p} = a + bL^2 + cW^2 + dH^2 \tag{A4}$$

where L, W and H denote the maximum length, width and thickness of anisotropic particles, respectively, while a, b, c and d are constants that can be determined by fitting experimentally derived SA_p with Eq. (A4) [4]. In PVA/NBC bionanocomposites, it is assumed that NBCs are uniformed dispersed within PVA matrices, resulting in two typical categories for the particle–matrix interaction including fully embedded and partially embedded NBCs within PVA matrices. Equation (A4) can be rewritten for calculating surface areas of outer interface ($\text{SA}_{\text{outer Interface}}$) and inner interface ($\text{SA}_{\text{inner Interface}}$) for a wide range of fully and partially embedded NBCs (subscripts of 'f' and 'p' mean fully and partially embedded NBCs) given by:

$$(\text{SA}_{\text{outer Interface}})_\text{f} = a_1 + b_1 L^2_{\text{Interphase}} + c_1 W^2_{\text{Interphase}} + d_1 H^2_{\text{Interphase}} \tag{A5}$$

$$(\text{SA}_{\text{outer Interface}})_\text{p} = a_2 + b_2 L'^2_{\text{Interphase-effective}} + c_2 W'^2_{\text{Interphase-effective}} + d_2 H'^2_{\text{Interphase-effective}} \tag{A6}$$

$$(\text{SA}_{\text{inner Interface}})_\text{f} = a_3 + b_3 L^2_{\text{NBC}} + c_3 W^2_{\text{NBC}} + d_3 H^2_{\text{NBC}} \tag{A7}$$

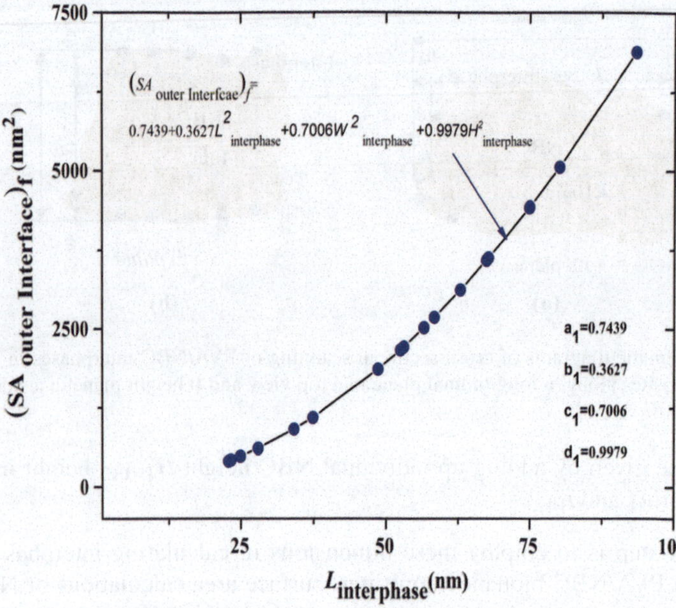

Fig. A3 Relationship between surface area of outer interface ($SA_{\text{outer Interphase}}$) and interphase length ($L_{\text{Interphase}}$) for PVA/NBC phases in PVA/NBC bionanocomposites [1]

$$(SA_{\text{inner Interface}})_p = a_4 + b_4 L'^2_{\text{NBC-effective}} + c_4 W'^2_{\text{NBC-effective}} + d_4 H'^2_{\text{NBC-effective}} \tag{A8}$$

where $L'_{\text{Interphase-effective}}$, $W'_{\text{Interphase-effective}}$ and $H'_{\text{Interphase-effective}}$ represent the maximum length, width and thickness for effective interphases. $L'_{\text{NBC-effective}}$, $W'_{\text{NBC-effective}}$ and $H'_{\text{NBC-effective}}$ denote the maximum length, width and thickness for effective NBCs. $a_1, b_1, c_1, d_1, a_2, b_2, c_2, d_2, a_3, b_3, c_3, d_3$ and a_4, b_4, c_4, d_4 are constants determined using curving fitting. For instance, the constants of a_1, b_1, c_1 and d_1 were determined to be 0.7439, 0.3627, 0.7006 and 0.9979, respectively, by fitting Eq. (A5) with measured surface area $(SA_{\text{outer Interface}})_f$ obtained from the AFM, as depicted in Fig. A3. Using such coefficients on the training set, the average error was about 8.7% while the maximum error is 26.7%, which were similar to those obtained from the validation sets.

- At the third step, the interphase volume was calculated based on the following equation suggested by Erdoğan [4]:

$$SA_p = eV_p^f \tag{A9}$$

where V represents the volume of anisotropic particles and e and f are also the constants to be determined by fitting experimentally acquired SA_p with

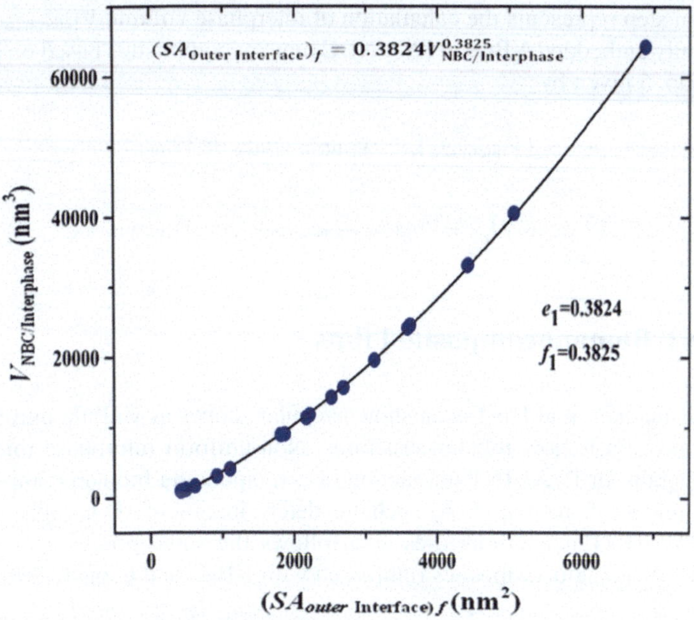

Fig. A4 Relationship between surface area of outer interface ($SA_{\text{outer Interphase}}$) and NBC/interphase volume ($V_{\text{NBC/Interphase}}$) in PVA/NBC bionanocomposites [1]

Eq. (A5). In case of PVA/NBC bionanocomposite system, Eq. (A9) can be rewritten to calculate interphase and NBC volume $V_{\text{NBC/Interphase}}$ as well as NBC volume V_{NBC} in case of fully and partially embedded NBCs by rearranging the following equations:

$$(\text{SA}_{\text{outer Interface}})_{\text{f}} = e_1 V^{f_1}_{\text{NBC/Interphase}} \tag{A10}$$

$$(\text{SA}_{\text{outer Interface}})_{\text{p}} = e_2 V^{f_2}_{\text{NBC/Interphase-effective}} \tag{A11}$$

$$(\text{SA}_{\text{inner Interface}})_{\text{f}} = e_3 V^{f_3}_{\text{NBC}} \tag{A12}$$

$$(\text{SA}_{\text{inner Interface}})_{\text{p}} = e_4 V^{f_4}_{\text{NBC-effective}} \tag{A13}$$

$e_1, f_1, e_2, f_2, e_3, f_3$ and e_4, f_4 are the constants determined using curving fitting. For example, the constants of e_1 and f_1 were estimated to be 0.3824 and 0.3825, respectively, with R^2 = ~0.995, as depicted in Fig. A4. Such an estimation was achieved by fitting Eq. (A10) with the data set of $V_{\text{NBC/Interphase}}$ being obtained from the AFM measurement.

- The final step represents the calculation of interphase volume $V_{\text{Interphase}}$ for fully or partially embedded NBCs in PVA/NBC bionanocomposites based on Fig. 5.1a, b in Chap. 5 given by

$$\left(V_{\text{Interphase}}\right)_{\text{f}} = V_{\text{NBC/Interphase}} - V_{\text{NBC}} \tag{A14}$$

$$\left(V_{\text{Interphase}}\right)_{\text{p}} = V_{\text{NBC/Interphase-effective}} - V_{\text{NBC-effective}} \tag{A15}$$

PVA/HNT Bionanocomposite Films

In a similar manner, real HNTs can show irregular shapes as well though they tend to mainly possess hollow tubular structures. Non-uniform interphase thickness is confirmed again for PVA/HNT interaction in corresponding bionanocomposites, as presented in Fig. 5.6 in Chap. 5. As such, the determination of 3D interphase dimensions of PVA/HNT nanocomposites also follows the same procedure mentioned earlier in PVA bionanocomposites reinforced with NBCs and Cloisite 30B clays as follows:

- Two typical diameters were assumed for PVA/HNT phases, as shown in Fig. A5a, consisting of the diameter along transverse section ($\text{DT}_{i\text{HNT}}$) and the diameter along longitudinal plane section ($\text{DL}_{j\text{HNT}}$). Consequently, to determine $W_{i\text{Interphase}}$ in PVA/HNT phases, they were scanned transversely, as depicted in Fig. A5 to yield:

$$W_{i\text{Interphase}} = \text{DT}_{i\text{HNT}} + t_{\text{Wi1}} + t_{\text{Wi2}} \tag{A16}$$

$L_{j\text{Interphase}}$ and $H_{k\text{Interphase}}$ of PVA/HNT phases were measured by scanning along the jth longitudinal plane ($j = 1, 2, 3, \ldots$) and along the kth height plane ($k = 1, 2, 3, \ldots$), as shown in Fig. A6a, b, respectively. Accordingly, Eqs. (A2) and (A3) can be rewritten as follows:

$$L_{j\text{Interphase}} = L_{j\text{HNT}} + t_{\text{Lj1}} + t_{\text{Lj2}} \tag{A17}$$

$$H_{k\text{Interphase}} = \text{DL}_{k\text{HNT}} + t_{\text{Hk1}} + t_{\text{Hk2}} \tag{A18}$$

- The next step is to determine interphase dimensions of PVA/HNT phases to calculate

$(\text{SA}_{\text{outer Interface}})_{\text{f}}$, $(\text{SA}_{\text{outer Interface}})_{\text{p}}$, $(\text{SA}_{\text{inner Interface}})_{\text{f}}$ and $(\text{SA}_{\text{inner Interface}})_{\text{p}}$ as given by:

$$(\text{SA}_{\text{outer Interface}})_{\text{f}} = a_1' + b_1' L^2_{\text{Interphase}} + c_1' W^2_{\text{Interphase}} + d_1' H^2_{\text{Interphase}} \tag{A19}$$

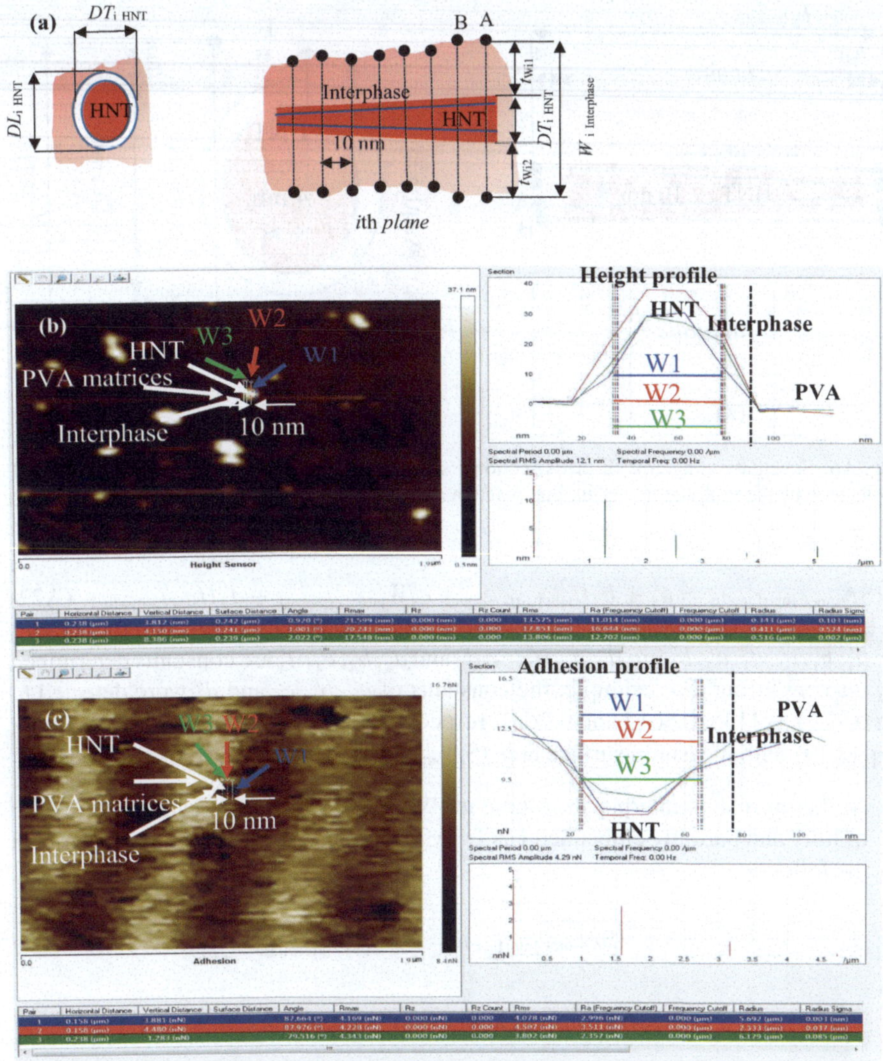

Fig. A5 **a** Schematic diagrams of cross-sectional scanning of PVA/HNT interphase in PVA/HNT bionanocomposites along transverse plane with **b** height profile and **c** adhesion profile [2]

$$(\mathrm{SA}_{\text{outer Interface}})_{\mathrm{p}} = a'_2 + b'_2 L'^2_{\text{Interphase-effective}} + c'_2 W'^2_{\text{Interphase-effective}} + d'_2 H'^2_{\text{Interphase-effective}} \quad \text{(A20)}$$

$$(\mathrm{SA}_{\text{inner Interface}})_{\mathrm{f}} = a'_3 + b'_3 L^2_{\mathrm{HNT}} + c'_3 W^2_{\mathrm{HNT}} + d'_3 H^2_{\mathrm{HNT}} \quad \text{(A21)}$$

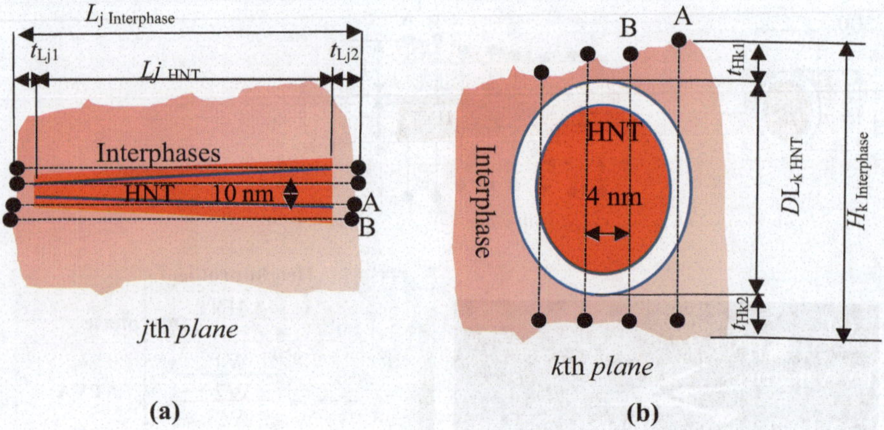

Fig. A6 Schematic diagrams of cross-sectional scanning of PVA/HNT interphase in PVA/HNT bionanocomposites along, **a** longitudinal plane in a top view, and **b** height plane in a side view [2]

$$(\text{SA}_{\text{inner Interface}})_{\text{p}} = a'_4 + b'_4 L'^2_{\text{HNT-effective}} + c'_4 W'^2_{\text{HNT-effective}} + d'_4 H'^2_{\text{HNT-effective}} \quad \text{(A22)}$$

$a'_1, b'_1, c'_1, d'_1, a'_2, b'_2, c'_2, d'_2, a'_3, b'_3, c'_3, d'_3$ and a'_4, b'_4, c'_4, d'_4 are constants determined by curving fitting. For example, the constants of a'_1, b'_1, c'_1 and d'_1 were detected to be 0.4932, 0.4135, 0.6613 and 0.7621, respectively, as depicted in Fig. A7 by fitting Eq. (A19) with measured surface area $(\text{SA}_{\text{outer Interface}})_{\text{f}}$ obtained from the AFM.

- At the third step, interphase volume in PVA/HNT phases was calculated in case of fully and partially embedded HNTs by rewriting Eqs. (5.5)–(5.8) in Chap. 5 in the following:

$$(\text{SA}_{\text{outer Interface}})_{\text{f}} = e'_1 V^{f'_1}_{\text{HNT/Interphase}} \quad \text{(A23)}$$

$$(\text{SA}_{\text{outer Interface}})_{\text{p}} = e'_2 V^{f'_2}_{\text{HNT/Interphase-effective}} \quad \text{(A24)}$$

$$(\text{SA}_{\text{inner Interface}})_{\text{f}} = e'_3 V^{f'_3}_{\text{HNT}} \quad \text{(A25)}$$

$$(\text{SA}_{\text{inner Interface}})_{\text{p}} = e'_4 V^{f'_4}_{\text{HNT-effective}} \quad \text{(A26)}$$

e'_1, f'_1, e'_2, f'_2 e'_3, f'_3 and e'_4, f'_4 are constants determined by curving fitting. For instance, the constants of e'_1 and f'_1 were calculated to be 0.4432 and 0.4726, respectively, with R^2 = ~0.95, as shown in Fig. A8.

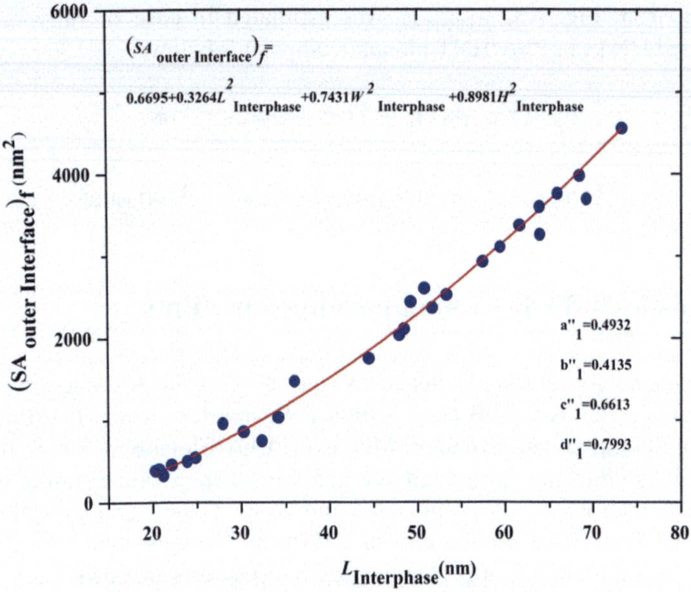

Fig. A7 Relationship between surface area of outer interface ($SA_{\text{outer interface}}$) and interphase length ($L_{\text{Interphase}}$) for PVA/HNT phases in PVA/HNT bionanocomposites [2]

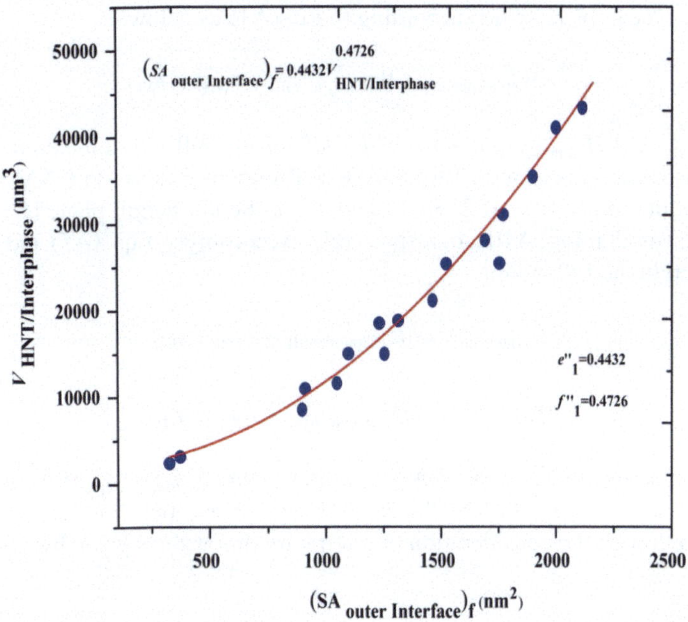

Fig. A8 Relationship between surface area of outer interface ($SA_{\text{outer Interphase}}$) and HNT/interphase volume ($V_{\text{HNT/Interphase}}$) in PVA/HNT bionanocomposites [2]

- As seen from Fig. A8, $V_{\text{Interphase}}$ was estimated in case of fully or partially embedded HNTs in PVA/HNT bionanocomposites below:

$$\left(V_{\text{Interphase}}\right)_{\text{f}} = V_{\text{HNT/Interphase}} - V_{\text{HNT}} \quad \text{(A27)}$$

$$\left(V_{\text{Interphase}}\right)_{\text{p}} = V_{\text{HNT/Interphase-effective}} - V_{\text{HNT-effective}} \quad \text{(A28)}$$

PVA/Cloisite 30B Clay Bionanocomposite Films

Despite a general concept of platelet-like shapes for Cloisite 30B clays, morphological structures of Cloisite 30B clays within PVA matrices reveal that their shapes can be more complex and irregular. Moreover, PVA/Cloisite 30B clay interphase has non-uniform thickness presented in Fig. 5.7 in Chap. 5 with a similar behaviour to PVA/NBC interphase in PVA bionanocomposites. Hence, an identical approach to detect 3D interphase dimensions in PVA/NBC bionanocomposites was also employed for PVA/Cloisite 30B clay bionanocomposites as follows:

- First of all, PVA/Cloisite 30B clay phases were scanned transversely in Fig. A9 in order to detect $W_{i\text{Interphase}}$. The scanning process was carried out within a size interval of 10 nm. Interphase width $W_{i\text{Interphase}}$ along the ith transverse plane (i = 1, 2, 3, …) can be rewritten according to Eq. (A1) as below:

$$W_{i\,\text{Interphase}} = W_{i\text{Cloisite 30B}} + t_{\text{W}i1} + t_{\text{W}i2} \quad \text{(A29)}$$

$L_{j\text{Interphase}}$ and $H_{k\text{Interphase}}$ in case of PVA/Cloisite 30B clay bionanocomposites were determined by scanned PVA/Cloisite 30B clay phases via PFQNM along the jth longitudinal plane (j = 1, 2, 3, …) and along the kth height plane (k = 1, 2, 3, …), as presented in Fig. A10a, b, respectively. Accordingly, Eqs. (A5) and (A6) can be rewritten in the following:

$$L_{j\text{Interphase}} = L_{j\text{Cloisite 30B}} + t_{\text{L}j1} + t_{\text{L}j2} \quad \text{(A30)}$$

$$H_{k\text{Interphase}} = H_{k\text{Cloisite 30B}} + t_{\text{H}k1} + t_{\text{H}k2} \quad \text{(A31)}$$

- The next step is to calculate $(\text{SA}_{\text{outer Interface}})_{\text{f}}$, $(\text{SA}_{\text{outer Interface}})_{\text{p}}$, $(\text{SA}_{\text{inner Interface}})_{\text{f}}$ and $(\text{SA}_{\text{inner Interface}})_{\text{p}}$ in PVA/Cloisite 30B clay phases by using Eqs. (5.1)–(5.4) in Chap. 5 with the consideration of interphase dimensions given by:

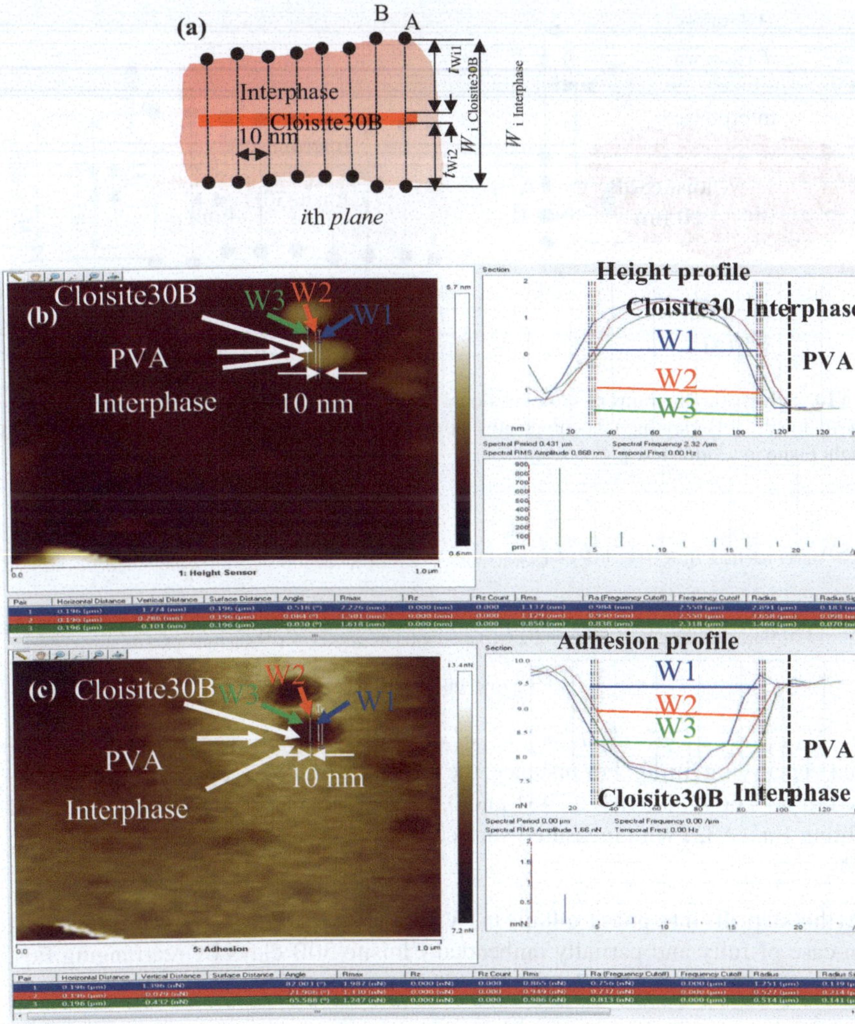

Fig. A9 **a** Schematic diagrams of cross-sectional scanning of PVA/Cloisite 30B clay interphase in PVA/Cloisite 30B clay bionanocomposites along transverse plane with **b** height profile and **c** adhesion profile [2]

$$(\mathrm{SA}_{\text{outer Interface}})_{\mathrm{f}} = a_1'' + b_1'' L_{\text{Interphase}}^2 + c_1'' W_{\text{Interphase}}^2 + d_1'' H_{\text{Interphase}}^2 \quad \text{(A32)}$$

$$(\mathrm{SA}_{\text{outer Interface}})_{\mathrm{p}} = a_2'' + b_2'' L_{\text{Interphase-effective}}'^2 + c_2'' W_{\text{Interphase-effective}}'^2 + d_2'' H_{\text{Interphase-effective}}'^2 \quad \text{(A33)}$$

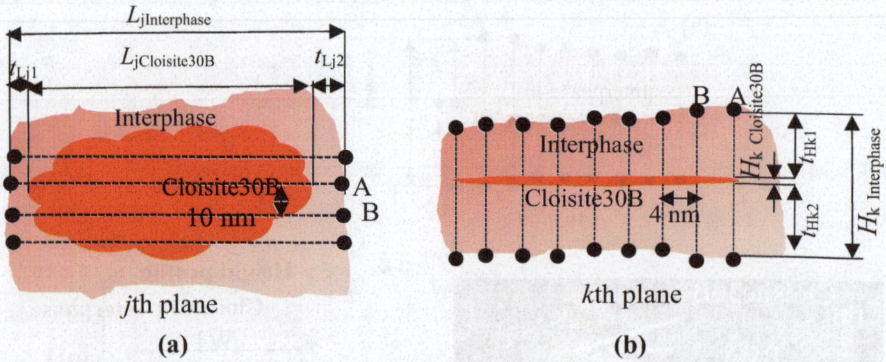

Fig. A10 Schematic diagrams of cross-sectional scanning of PVA/Cloisite 30B clay interphases in PVA/Cloisite 30B clay bionanocomposites along, **a** longitudinal plane both in a top view, and **b** height plane in a side view [2]

$$(\text{SA}_{\text{inner Interface}})_{\text{f}} = a_3'' + b_3'' L^2_{\text{Cloisite 30B}} + c_3'' W^2_{\text{Cloisite 30B}} + d_3'' H^2_{\text{Cloisite 30B}} \quad \text{(A34)}$$

$$(\text{SA}_{\text{inner Interface}})_{\text{p}} = a_4'' + b_4'' L'^2_{\text{Cloisite 30B-effective}} + c_4'' W'^2_{\text{Cloisite 30B-effective}} + d_4'' H'^2_{\text{Cloisite 30B-effective}} \quad \text{(A35)}$$

a_1'', b_1'', c_1'', d_1'', a_2'', b_2'', c_2'', d_2'', a_3'', b_3'', c_3'', d_3'' and a_4'', b_4'', c_4'', d_4'' are constants determined via curving fitting. For instance, the constants of a_1'', b_1'', c_1'' and d_1'' were determined to be 0.6695, 0.3264, 0.7431 and 0.8981, respectively, as shown in Fig. A11 by fitting Eq. (A32) with measured surface area $(\text{SA}_{\text{outer Interface}})_{\text{f}}$ obtained from the AFM.

- At this step, the interphase volume in PVA/Cloisite 30B clay phases was calculated in case of fully and partially embedded Cloisite 30B clays by rearranging Eqs. (5.5)–(5.8) in Chap. 5 to yield:

$$(\text{SA}_{\text{outer Interface}})_{\text{f}} = e_1'' V^{f_1''}_{\text{Cloisite 30B/Interphase}} \quad \text{(A36)}$$

$$(\text{SA}_{\text{outer Interface}})_{\text{p}} = e_2'' V^{f_2''}_{\text{Cloisite 30B/Interphase-effective}} \quad \text{(A37)}$$

$$(\text{SA}_{\text{inner Interface}})_{\text{f}} = e_3'' V^{f_3''}_{\text{Cloisite 30B}} \quad \text{(A38)}$$

$$(\text{SA}_{\text{inner Interface}})_{\text{p}} = e_4'' V^{f_4''}_{\text{Cloisite 30B-effective}} \quad \text{(A39)}$$

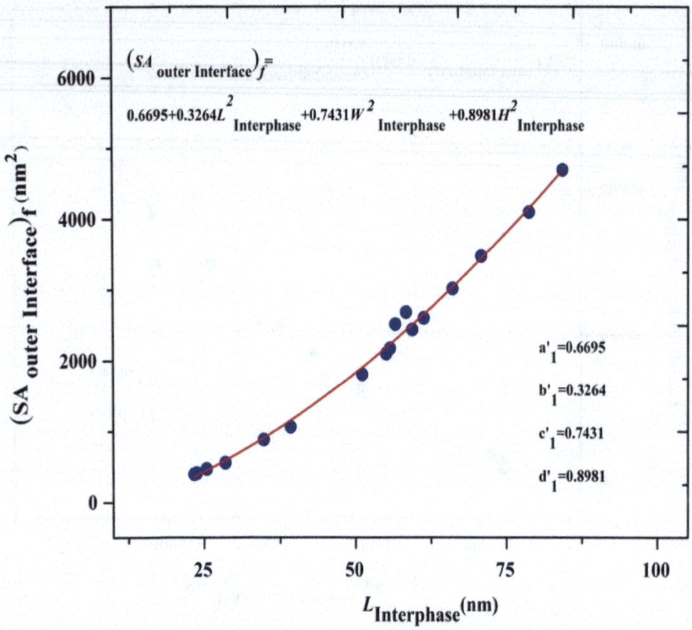

Fig. A11 Relationship between surface area of outer interface ($SA_{\text{outer interface}}$) and interphase length ($L_{\text{Interphase}}$) for PVA/Cloisite 30B clay phases in PVA/Cloisite 30B clay bionanocomposites [2]

e''_1, f''_1, e''_2, f''_2, e''_3, f''_3 and e''_4, f''_4 are constants determined using curving fitting. For instance, the constants of e''_1 and f''_1 were calculated to be 0.4015 and 0.4116, respectively, with $R^2 = {\sim}0.96$, as shown in Fig. A12.

- At the last step, $V_{\text{Interphase}}$ was calculated for fully or partially embedded Cloisite 30B clays in PVA/ Cloisite 30B clay bionanocomposites based on Fig. 5.1c, d in Chap. 5, which is given by

$$(V_{\text{Interphase}})_{\text{f}} = V_{\text{Cloisite 30B/Interphase}} - V_{\text{Cloisite 30B}} \tag{A40}$$

$$\left(V_{\text{Interphase}}\right)_{\text{p}} = V_{\text{Cloisite 30B/Interphase-effective}} - V_{\text{Cloisite 30B-effective}} \tag{A41}$$

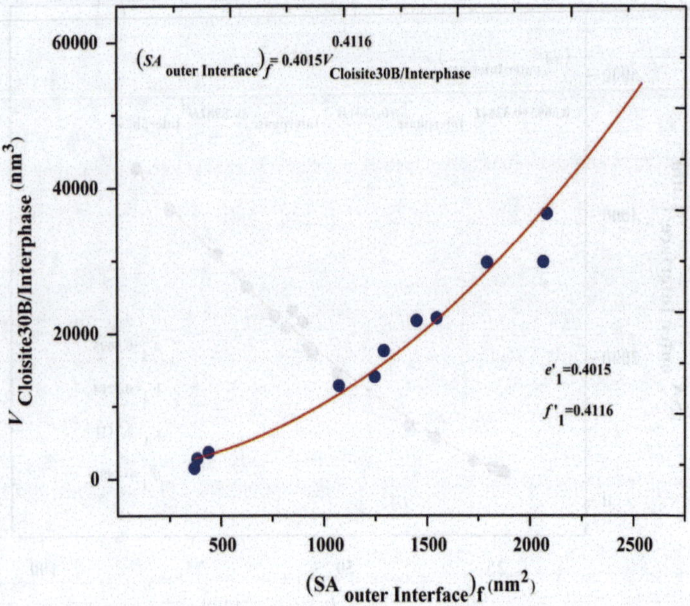

Fig. A12 Relationship between surface area of outer interface ($SA_{outer\ interface}$) and Cloisite 30B clay/interphase volume ($V_{Cloisite\ 30B/Interphase}$) in PVA/Cloisite 30B clay bionanocomposites [2]

References

1. Mousa M, Dong Y (2018) Novel three-dimensional interphase characterisation of polymer nanocomposites using nanoscaled topography. Nanotechnology 29(38):385701
2. Mousa M, Dong Y (2020) Towards sophisticated 3D interphase modelling of advanced bionanocomposites via atomic force microscopy. J Nanomater Article ID 4526108
3. Behmer DJ, Hawkins CP (1986) Effects of overhead canopy on macroinvertebrate production in a Utah stream. Freshw Biol 16(3):287–300
4. Erdoğan S (2016) Simple estimation of the surface area of irregular 3D particles. J Mater Civ Eng 28(8):04016062

M. Mousa and Y. Dong, *Multiscaled PVA Bionanocomposite Films*,
https://doi.org/10.1007/978-981-15-8771-9

The manufacturer's authorised representative in the EU is Springer Nature Customer Service Centre GmbH, Europaplatz 3, 69115 Heidelberg, Germany. If you have any concerns regarding our products, please contact ProductSafety@springernature.com

Printed and bound by CPI Group (UK) Ltd, Croydon, CR0 4YY
06/10/2026
02225252-0001